Save yourself the time and trouble of manual file entry...

Order the
Fractal Programming in Turbo Pascal Disk

This optional disk contains all the individual programs listed in the book. Source code for over 100 black and white pictures and over 50 color pictures are provided, allowing you to reproduce them. MS-DOS format. Requires PC or clone with EGA or VGA and color monitor; and Turbo Pascal 4.0 or later.

To order, simply return this postage-paid card with your payment to: M&T Books, 501 Galveston Drive, Redwood City, CA 94063-4728. Or, **call toll-free 800-533-4372 (In CA 800-356-2002)**. Also available in 3 1/2" format.

YES! Please send me the *Fractal Programming in Turbo Pascal* Disk for $20_____

California residents add applicable sales tax ____ % _____

TOTAL _____

Check enclosed. Make payable to **M&T Books**.

Charge my ____ VISA ____ MasterCard ____ American Express

Card # _____ Exp. date _____

Name _____

Address _____

City _____ State _____ Zip _____

Note: Disk may be exchanged for a replacement disk if damaged. Disks may not be returned for credit or refund.   7065

BUSINESS REPLY MAIL
FIRST CLASS PERMIT 871 REDWOOD CITY, CA

POSTAGE WILL BE PAID BY ADDRESSEE

M&T BOOKS

501 Galveston Drive
Redwood City, CA 94063

PLEASE FOLD ALONG LINE AND STAPLE OR TAPE CLOSED

Fractal
Programming in Turbo Pascal

M&T BOOKS

Fractal
Programming in Turbo Pascal

Roger T. Stevens

M&T BOOKS

M&T Publishing, Inc.
Redwood City, California

M&T Books

A Division of M&T Publishing, Inc.
501 Galveston Drive
Redwood City, CA 94063

M&T Books

General Manager, Linda Hanger
Acquisitions Editor, Brenda McLaughlin
Operations Manager, Michelle Hudun
Technical Editor, Scott Robert Ladd
Senior Project Editor, David Rosenthal
Assistant Project Editor, Kurt Rosenthal
Cover Art Designer, Michael Hollister

Library of Congress Cataloging Card Number

90-5471

ISBN 1-55851-106-7   (book/disk) $39.95
ISBN 1-55851-107-5   (book) $24.95
ISBN 1-55851-108-3   (disk) $20.00

94 93 92 91 90   5 4 3 2 1

Acknowledgments

All of the software in this book was written in Turbo Pascal version 5.5 furnished by Borland International, 4385 Scotts Valley Drive, Scotts Valley, CA 95066.

Valuable technical information on the format of .PCX files and a copy of PC Paintbrush were supplied by Shannon of Z-Soft Corporation, 1950 Spectrum Circle, Marietta, GA 30067.

Dr. Michael Batty of the University of Wales Institute of Science and Technology was kind enough to send me several reprints of his publications and direct me to his book *Microcomputer Graphics*, which contains much useful information.

All software was checked out on a computer having a Vega VGA card furnished by Video Seven, Inc., 46335 Landing Parkway, Fremont, CA 94538, and a NEC Multisync Plus Color Monitor furnished by NEC Home Electronics (U. S. A.) Inc.

Limits of Liability
and Disclaimer of Warranty

How to Order
the Accompanying Disk

Fractal Programming in Turbo Pascal is a comprehensive "how to" book written for programmers interested in fractals. Learn about reproducing those developments that have changed our thinking about the physical sciences, and create pictures that have both beauty and an underlying mathematical meaning. Included are more than 100 black-and-white pictures and 32 color pictures. All source code to reproduce these pictures is provided on disk, in MS-DOS format. It requires a PC or clone with an EGA or VGA card and a color monitor; and Turbo Pascal 4.0 or later.

The disk is $20, plus sales tax if you are a California resident. Order by sending a check, or credit card number and expiration date, to:

Fractal Programming in Turbo Pascal Disk
M&T Books
501 Galveston Drive
Redwood City, CA 94063

Or, you may order by calling our toll-free number between 8 A.M. and 5:00 P.M. Pacific Standard Time: 800/533-4372 (800/356-2002 in California).

Contents

1

Introduction

Alfred North Whitehead, the American mathematician and philosopher, was fond of relating how physicists, at the end of the nineteenth century, considered physics to be essentially a closed book. Everything of any importance in the field was already known. All that needed to be done was to clean up a few loose ends and the volume could be marked "COMPLETE" and closed forever. Then, in the course of cleaning up the loose ends, Schrödinger discovered quantum mechanics and Einstein created the theory of relativity and physics was in a state of revolution, with more unanswered questions than ever before. To Whitehead, one of the most significant aspects of this revolution was its effect upon the philosophical outlook of physicists. Never again would they take the smug, self-contained approach that everything was known and complete. Instead, their minds would always be open to the myriad of possibilities of the unknown.

By 1980, however, things had gone full circle, and the cosmologist extraordinaire Stephen Hawking presented a lecture, "Is the End in Sight for Theoretical Physics?", in which he postulated that we already know everything about physics that is important in daily life, and that future discoveries would require huge sums of money and large machines to discover insignificant refinements. While Hawking was closing out physics, a revolution in scientific thinking had already begun which would cut across disciplinary boundaries so that physics, as well as other sciences, would never be the same again. The name of this revolution is *Chaos*.

For centuries, mathematicians were comfortable with an intuitive feeling for what might happen when they wrote down systems of equations. A simple set of equations would produce simple results. In most cases, the end result would be a simple, stable expression that represented the end state of the system. If things got a little more complicated, the equation might blow up, meaning that there were unfortunate sets of inputs for which the result would go off toward infinity. In other situations, the result might be a periodic function, which would never reach an end value, but would at least settle down to a regular repeating function that could be easily predicted. In the real world, situations existed where the state of a system could not be predicted at any given time. Mathematicians got around this problem by representing the system state through the selection of random numbers. They often referred to the system as *noisy*, where noise was a function that took on completely random values, over a given range, through time. The intuitive feeling was that noise represented the results of some regular functions that we did not yet know how to define and measure and that as soon as our understanding and methods improved a little more, we could fully understand, characterize, and eliminate if necessary, the effects of noise.

"Monster" Curves

The first cracks in this structure began to appear in the late nineteenth and early twentieth century, when mathematicians such as Cantor, von Koch, and Peano began to draw curves quite unlike those that mathematicians had ever seen before. They were often undifferentiable. They were usually self-similar (the shape of each small segment of the curve was the same as the shape of a much larger segment), their length could not easily be measured or defined, and their dimension appeared to differ from the traditional dimension of one for a line and to perhaps be somewhere between a line and a plane.

Traditional mathematicians called these curves "monsters" and "pathological" and refused to deal with them at all. Lacking the tools of modern computers, not much progress was made in studying these curves for many years. In chapters 6, 7, 8, and 9, we shall look at some of these curves in considerable detail and provide software for drawing and investigating them.

Strange Attractors

The chaos revolution really began in about 1961. Edward Lorenz, at the Massachusetts Institute of Technology, was attempting to develop a model for weather systems which would make improved weather forecasting possible. His model appeared to be a fairly good representation of weather patterns, which when run produced results similar to the kind of weather that actually occurred.

One day, Lorenz wanted to pick up from the middle of a previous computer run and examine a sequence in greater detail. He typed in his intermediate data and started the computer again. To his dismay, the new computer run started by duplicating the results of the previous one, but then began to diverge farther and farther. Lorenz satisfied himself that these results were not due to a faulty computer, and ultimately determined that the cause was that he had typed in the intermediate results to only three decimal places, whereas the computer had originally stored them to six decimal places.

This appeared to be bad news for weather forecasters; if over a period of weeks weather patterns could be completely different due to differences in the fourth or higher decimal places of input data, there appeared to be little possibility that forecasters could collect accurate enough data to make accurate long range forecasting possible. Lorenz eventually boiled down his model to three simple differential equations, which also happened to represent fluid flow or the action of a particular type of water wheel. The result of these equations, over time, was not a single stable result or a periodic function. But it was not random noise either. Instead, a curve appeared which was ordered and predictable, but never the same. Basically, regardless of input, this set of equations settled down to values from within a family of curves. Fortunately the curves took on a set of predictable values; unfortunately, the curves continued on to infinity without ever repeating themselves. These curves became known as the Lorenz attractor. It was the first of the strange attractors. The Lorenz equations, a program for graphing the Lorenz attractor, and equations and software for other strange attractors are given in Chapter 4.

Population Curves and Bifurcation Diagrams

In the early 1970s, Robert May, at the Institute for Advanced Studies at Princeton, was looking at the mathematics of population growth. The critical equation was:

$$x_n = rx_{n-1}(1 - x_{n-1}) \qquad \text{(Equation 1-1)}$$

This simple equation had been assumed to have two outcomes: either a population achieved a stable equilibrium value or it tapered off to extinction. As May experimented with different values of the parameter r, however, a strange phenomena occurred. As the parameter grew larger, the result ceased to achieve a stable equilibrium and instead began to oscillate between two different values. A little larger value of the parameter and there were four alternating stable states, then eight, and so forth until the behavior became chaotic and didn't settle down to any value at all. But then, as the parameter increased some more, a stable window was found in the middle of chaos, with three alternating states which then increased to six, twelve, and finally back to chaos again. Another window, farther on, began with seven alternating states.

May's friend James Yorke at the Institute for Physical Science and Technology at the University of Maryland, did a rigorous mathematical analysis of the behavior of this equation and proved that if a regular cycle of period three ever occurs in any one dimensional system, then the same system will also display regular cycles of every other possible length and various completely chaotic cycles as well.

Yorke and Tien-Yien Li wrote a paper on this, which was mischievously called *Period Three Implies Chaos*. This is the origin for the name *Chaos* for this new field of science. A few years later, Mitchell Feigenbaum was studying the same equation at the Los Alamos National Laboratory. He observed a regularity in the period doubling effect, which had a ratio of 4.6692016090, now known as the *Feigenbaum number*. Strangely enough, this same ratio applies to period doubling in a wide variety of iterated equations; almost any iterated equation for which the basic equation produces a curve with a hump. Software to produce bifurcation diagrams and investigate the Feigenbaum number is found in Chapter 5.

Mandelbrot and Julia Sets

At about this same time, Benoit Mandelbrot at IBM's Thomas J. Watson Research Center was taking a closer look at the von Koch and Peano curves. A technique had been developed years before for assigning a dimension greater than the standard Euclidian dimension to such curves. This dimension is known as the *Hausdorff-Besicovitch* dimension. Mandelbrot coined the term *fractals* to describe all curves whose Hausdorff-Besicovitch dimension is greater than their Euclidian dimension.

Mandelbrot was also looking at the characteristics of Julia sets, an intriguing variety of curves based upon mapping the function:

$$z_n = z_{n-1}^2 + c \qquad\qquad \text{(Equation 1-2)}$$

FRACTAL PROGRAMMING IN TURBO PASCAL

where z and c are complex numbers. Mandelbrot developed a new way of mapping this equation: the *Mandelbrot set*. This set also turned out to be a kind of catalog of all possible Julia sets, from which particularly interesting Julia set parameters may be selected for mapping. Mandelbrot was beginning to discover the same characteristic discovered by Lorenz: that very simple mathematical expressions can result in chaotic non-periodic functions, which nonetheless do have a very rigid kind of order that is completely specified by the original equations. A complete discussion of the Mandelbrot set, together with software for plotting and investigating it, is given in Chapter 12. Chapter 13 describes in detail how to create displays of the Julia sets.

Mandelbrot began to develop an intuition, which has proved to be right in many cases, that fractals are the natural way of representing many of the shapes in nature. Thus, just as Euclidian geometry is the natural way of describing man-made shapes such as squares, triangles or cubes, so fractals are the natural language for describing clouds, trees, leaves, and other natural objects. This seems to make sense, since we know that apparently very complex natural objects often are generated from rather limited genetic codes.

Iterated Function Systems

Michael Barnsley investigated Julia sets, and looked for ways to produce even more variability and, perhaps, to generate patterns that matched those of living things. Barnsley discovered what he called *iterated function systems*. Basically, such a system consisted of several sets of equations, each of which represents a rotation, a translation, and a scaling. By starting with a point and randomly applying one of his sets of equations, according to specified probability rules, Barnsley could generate classic fractals, and he soon discovered how to make the rules for generating ferns and other shapes from nature. Chapter 20 provides a description of this technique and some software for generating various shapes.

The State of Science

Until scientists, equipped with the capabilities of modern computers, began investigating the characteristics of iterated equations, it was assumed that a simple equation produced a simple result and a more complex equation a more complex result. Investigators delving into either mathematics or physical sciences looked for well-behaved functions and tended to ignore or bypass non-linear effects. The idea that a very simple expression could produce complex, non- periodic, but regular behavior had not been conceived. Evidences of these effects were passed off as "noise" or as "experimental error."

Today, the effects of investigation into chaos and the application of fractals are changing the way we think about many aspects of the physical sciences and are opening up new areas in mathematics. The news for science is good and bad. The bad news is that things are much more complicated than we thought. The good news is that things are much simpler than we thought. To expand this thought a little, the good news is that many very complex structures and very complex behaviors can be expressed by very simple iterated equations. The bad news is that the structures and behaviors are aperiodic and that portions of the curve which are far apart, travelling along the curve, can be physically located so close together that absolutely precise knowledge of coordinates is necessary to know which portion of the curve we are on. We can only predict the future of the curve accurately, if we know exactly where we are on it, and this requires more precise measurement of our present position than we are capable of making.

Why this Book is Different

One might guess, from the brief survey given above, that the introduction of the new field of chaos into the sciences was not greeted with tremendous enthusiasm by many scientists. There is a myth that scientists are totally objective, impassionately conducting experiments and using the results to discover truth. Actually, scientists are persons, not unlike you and me. Their investigations and theories are often directed by prejudice, and the "truth" that they come up with is often only one truth from many and that truth the one which their predisposition has lead them to discover. The result, as far as chaos is concerned, is that those who were pioneers in the field tended to be a particular type of person. They needed to be a unique combination of scientist, philosopher, and artist, with a reasonable amount of stubbornness and a little eccentricity thrown in. The result is that many of the books which are currently available on fractals bear some resemblance to treatises on medieval alchemy.

These books are filled with esoteric equations and beautiful illustrations of results, but the mechanics of how to get from one to the other is slighted or missing altogether. This book takes a completely different approach. It's purpose is to provide you with software that you can use to duplicate many of the fractal pictures and basic diagrams of chaos and to proceed from there to easily modify the software to whatever new results come from your own ideas. If you have seen some of the beautiful pictures produced by fractal programs and want to produce the same kind of pictures on your IBM PC (or clone) or perhaps create new and interesting pictures that have never before been seen, or if you are interested in using the basic tools to apply fractals to problems in the physical sciences, then this book is for you. It discusses all of the well-known types of fractal curves and provides Turbo Pascal language programs to reproduce all of the pictures that are shown in this book. Hints are given on how to modify parameters so as to create your own original pictures.

This book is a companion to the earlier volume, Fractal Programming in C, which has basically the same content, but gives programs in the C language. If you are already familiar with Turbo Pascal and don't want to learn C, you can use this book to generate fractal programs. This book uses the Turbo Pascal graphics routines whenever possible. Appendix A provides some special purpose programs, including those to save and restore graphics screens. All of the programs were written using Turbo Pascal version 5.5, but since none of the new object-oriented features were used, they should work equally well with Turbo Pascal versions 4.0 and 5.0. These are the tools; how you use them depends strictly on your skill and imagination. The whole field of chaos is still new. There is plenty of room for new discoveries or new art. Good luck!

2

What Are Fractals?

When I tell people that I have been writing a book on fractals, they usually respond with two questions. The first is "What are fractals?" and the second is "What are fractals good for?" If I am feeling in an ornery mode, I respond to the first question with Mandelbrot's classical definition: "A fractal is a curve whose Hausdorff-Besicovitch dimension is larger than its Euclidian dimension." But more than this is really required in explaining fractals, so let's start at the beginning.

The Beginning of Fractal Curves

Draw a line on a sheet of paper. Euclidean geometry tells us that this is a figure of one dimension, namely length. Now extend the line. Make it wind around and around, back and forth, without crossing itself, until it fills the entire sheet of paper. Euclidean geometry says that this is still a line, a figure of one dimension. But our intuition strongly tells us that if the line completely fills the entire plane, it must be two-dimensional.

Such thinking started a revolution in mathematics about a hundred years ago. Mathematicians such as Cantor, von Koch, Peano, Hausdorff, and Besicovitch drew curves that were called "monsters," "psychotic," and "pathological" by traditional mathematicians. A new type of dimensioning was proposed, in which a curve could have a fractional dimension, not just an integer one. Recursive techniques and iterated expressions were found that could describe curves that have fractional dimensions. But without high speed digital computers, the actual drawing of such curves was a long and tedious process. So little progress was made in this unusual field for nearly a hundred years.

The advent of digital computers made the investigation of such curves a fruitful field. From the early investigations, we could understand what we were trying to do. We wanted to draw curves that appeared to have more complex dimensional characteristics than were explained by traditional geometry. Computers were turned loose on very simple mathematical iterated expressions, in which the next state of a parameter depends solely on a simple relationship to the current state of the parameter. The iteration was performed many times and the resulting location of the parameter at each state was plotted. The resulting plots turned out to have many interesting characteristics. For one thing, they never repeated themselves. Furthermore, they tended to have the characteristic of self-similarity. In other words, if a small portion of the plot was enlarged, its shape was very much like a large portion of the original plot. Finally, the plots turned out to have shapes of great interest and extreme beauty.

The curves still didn't make much sense in terms of traditional mathematics, and consequently remained an anathema to traditional mathematicians. Dr. Benoit Mandelbrot was the first person to make use of a digital computer to investigate fractals in depth, and his results were not welcomed warmly by traditional mathematicians.

How are Fractals Used?

Explaining the use of fractals is a little more difficult. Mandelbrot contends that just as the shapes of traditional geometry are the natural way of representing man-made objects (squares, circles, triangles, etc.), fractal curves are the natural way of representing objects that occur in nature. Thus fractals have a value both as art objects and as a means of representing natural scenes. Moreover, fractals occur naturally in the expressions for mathematical phenomena as varied as the prediction of weather systems, the describing of turbulent flow of liquid, and the growth and decline of populations. Finally, fractals are useful in dimensional transformations that can be used for expressing and compressing graphical data. Ignoring the artistic value, the best answer to the question, "What are fractals good for?" is the reply, "Fractals appear to provide solutions to many previously unanswered questions at the frontiers of the physical sciences." Consequently, to work at the frontiers of science, one needs to understand what fractals are and how to work with them.

In the later chapters of this book, we shall attempt to do our own experimenting with the creation and modification of fractal curves. We shall not spend too much time worrying about the uses of fractals in the sciences, but will concentrate on understanding as many different types of fractal curves as possible and developing computer programs to generate these curves. Then, when we encounter a physical problem that requires a fractal solution, we will know what to do and how to do it.

Basic Considerations

Let's establish some points of orientation that will be useful in practical investigations of the chaotic field of fractals. First, intuition leads us to believe that fractal curves should have a dimension greater than their traditional geometric dimension. Second, there is now sound mathematical grounding for accepting this premise. Third, fractal curves are associated with many physical and natural phenomena. Fourth, fractals often possess a rare and unusual beauty. No doubt, this is partly true because fractals correspond to the way in which nature produces those shapes that we are most familiar with and that basically define our ideas of "the beautiful." Finally, fractals have the unusual characteristic that they can be defined totally by relatively simple mathematical equations, yet they are not periodic.

Thus the progression of the fractal curve differs widely if we start at just slightly different points in space, so unless we can measure where we are with absolute precision, we cannot be sure just what the progression of the curve will look like. This is in spite of the fact that the curve is defined through all of its wanderings by very simple iterated expressions. Most fractals are self-similar, so that the shape that we identify in the plot of a fractal curve repeats itself on a smaller and smaller scale as we enlarge the image further and further.

Fractal Dimensions

Let's return to the statement made at the beginning of this chapter that a fractal is a curve whose Hausdorff-Besicovitch dimension is greater than its Euclidian dimension. We now have some idea of the nature of fractal curves and of what this new definition of dimension means, but that doesn't help much unless we can actually come up with some meaningful dimensional numbers. A rigorous definition of the Hausdorff-Besicovitch dimension is a rather lengthy mathematical process, and for many fractals it is almost impossible to determine this dimension. However, for a large class of self-similar fractals, which we will discuss in Chapters 6, 7, and 8, the fractal dimension is easily obtained.

Suppose that we start with an initiator that is some simple geometric figure consisting of a number of connected line segments. It may be a triangle or a square or even just a straight line. We now define a generator. This generator is is a series of line segments which is going to replace every line segment of the initiator. The generator consists of N line segments, each of length r, where r is a fraction of the line segment being replaced. The arrangement of the N line segments is such that the distance from the beginning of the generator to its end is the same as the length of the line segment being replaced. The replacement process repeats an infinite number of times, each time replacing each line segment of the previous level curve with a scaled down replica of the generator. It can then be shown that the Hausdorff-Besicovitch dimension of the resulting fractal curve is:

$$D = \log N \ / \ \log(1/r) \qquad \text{(Equation 2-1)}$$

Comparing this dimension with the Euclidian dimension gives us some idea of the properties of a fractal. For example, a D of 1.0 is simply an ordinary line, whereas a D of 2.0 means that the curve completely fills the plane.

Background Material

With this background in mind, let's begin looking at some fractal curves and creating some software to view them and work with them on our IBM PC computers. But first, we need to develop some special purpose tools that are needed for the generation of fractals. Whenever possible, we shall use the graphics functions that are provided as a part of the Turbo Pascal package. But sometimes, these just won't do the job properly, or we want some unique capability that is beyond the bounds of Turbo Pascal. The next chapter describes a package of such routines and describes how to incorporate them in your Turbo Pascal programs. If all goes well, this is all you need to know. Unfortunately, there are more hardware and software combinations in existence than I can conceive of. If you have one that I haven't considered, and as a result something goes wrong, you may want to know something about the baseline systems that I used to develop the software. Therefore, those systems are described in Appendix A. If you appear to be having an incompatibility problem between the program listings and your hardware, take a close look at the hardware configurations listed and try to determine how yours differs from the norm. Once these preliminaries are out of the way, we will go ahead and generate some fractal displays.

3

Tools for Creating Fractals

The Turbo Pascal graphics package contains most of the functions and procedures that are needed for the generation of fractal curves. However, there are a few things that can't be handled by the standard graphics package. In addition there are procedures for saving a display screen to a disk file and for restoring the screen from the disk file and functions and procedures for using turtle graphics in a modified form. All of these are used on numerous occasions throughout the book. It seems most appropriate to discuss these routines in a single central place, namely this chapter. If you're not too familiar with Turbo Pascal 5.5, you can learn a lot about the language by carefully reading the descriptions that follow and looking at the program listing.

None of the programs to be described are very large, so if you want to take the simplest way out, you can just add those that are needed to each fractal program as you create it. However, Turbo Pascal supports the concept of units, which permits convenient grouping of frequently used functions and procedures. We shall discuss units in detail further in the chapter. First, however, we'll describe each of the individual functions and procedures that make up the package. All of these functions and procedures are listed in the implementation section of the complete fractal unit listing given in Figure 3-2. (For Turbo Pascal, a function is a subroutine that returns a value; a procedure is a subroutine that does not return a value.)

Setting the EGA Palettes

The EGA permits us to set each of the sixteen different colors (referred to as *palettes*) that can be displayed on the screen at one time to any of the 64 color shades available with the EGA when in high resolution mode 16. For some strange reason, the Turbo Pascal procedure for setting the EGA palette only permits setting of the 16 palettes to any of the 16 default colors rather than to any of the available 64 colors. For many of our fractal displays, this is an unacceptable limitation, so we shall make use of our own procedure for setting the palette.

Figure 3-1 shows the way in which a color number is stored within an EGA palette register. In the case of the default colors, the low intensity set of colors corresponds to a 75% level of the colors specified by the bits set. The high intensity colors have both the 75% and 25% bits set for the selected colors, giving 100% of that color's level. There is one exception: low intensity yellow (brown), which is palette number 6, is set to 20H, which sets green to 25% and red to 75%.

Figure 3-1. Contents of Color Byte Sent to Palette Register

Bit	7	6	5	4	3	2	1	0
			25% Red	25% Green	25% Blue	75% Red	75% Green	75% Blue

As you can see, there are a lot of different shades not included in the defaults. The VGA operates in a quite different fashion, but for mode 16 it can be quite transparent and appear to be the same as the EGA. The VGA has 256 color registers, each of which can be set to 256K color hues. These registers are broken up into four sets of 64 registers. Any one of these four sets can be used to define the 64 colors from which the display selects the 16 palettes. When you send a number from 0 to 63 to the setEGApalette function while using the VGA, you are actually selecting one of the 64 registers from which the VGA takes the designated color hue. In the default condition, however, one set of 64 registers is always selected, and that set contains the same 64 shades of color that are designated by the numbers 0 to 63 by the EGA. Therefore, although the mechanism is quite different, sending a number to the setEGApalette function for the VGA results in the same color that is called up by using that function with the EGA. You can develop functions that choose a different set of 64 registers to define the colors and that change the color shadings of the registers from the default values, which makes it possible for you to access the highly increased number of color hues available on the VGA. Note, however, that for mode 16, you are always limited to 16 different colors on a display at one time.

The procedure to set the EGA palettes is included in Figure 3-2. It makes use of the ROM BIOS services to perform this task. Turbo Pascal, in its DOS unit, defines a type Registers, which contains variable names for all of the registers whose contents is to be transferred during an interrupt call. The color that is to be sent to a palette register is stored in the proper member of the PALETTE array, and then the registers are loaded with the proper values to cause the ROM BIOS to set the selected palette register to the selected color. The video services interrupt (10H) is then performed to set the palette register.

Note that since whatever value is sent to a palette register is also stored in the global array PALETTE, we can keep track of which colors have been set into the palette registers, even though this data is not recoverable from the EGA hardware. Thus we can use the contents of PALETTE to obtain color data to store in a file when we are saving a display and be able to regenerate the display with the same colors that were originally specified. However, the price that we pay for this is that we must include PALETTE as a global array in each of our programs that makes use of the setEGApalette function or as global data in the unit containing this procedure. If we don't do this, we will not be able to compile and run the program properly. Furthermore, if we are going to use the *save_screen* function, we must initialize the

PALETTE array with the proper default colors in order to have the correct color information for those colors that are not changed so that it can be saved to the file. The only exception to this initialization requirement is when we are going to set every one of the 16 EGA palettes to a new color.

Reading a Byte from a Color Plane

In the course of saving a display screen to a disk file, we will make use of a technique that reads the screen information a byte at a time from a single color plane (more about this later). We do, however, need a function (not available in the Turbo Pascal library) that will enable us to specify a pixel address (as an offset from the basic page address) and a color plane number and return a byte that represents the values of the specified pixel and the seven pixels following it for the specified color plane. The EGA does not permit obtaining this information directly from memory. Instead, you must set some of the EGA control registers so that when memory is read, only data from the specified color plane is allowed to be transferred. The registers are set using Turbo Pascal's Port procedure. Then the desired data is read directly from the proper memory address. The function is *read_screen* and is included in Figure 3-2.

Displaying a Byte on the Screen

The display function included in Figure 3-2 is just the reverse of the *read_screen* function just described. In other words, it takes an address offset, a color plane designator, and a byte of data and sends the data to the proper color plane of the display memory at the specified address. To do this with the EGA, we must first do a dummy read of the selected memory address to transfer the existing memory data to the internal registers of the EGA. Then we properly set the EGA registers with Turbo Pascal's Port procedure so that when we write the data byte to memory it only goes to the proper memory plane.

If you have a burning desire to learn just how the registers of the EGA and VGA work, get a copy of my book *Graphics Programming in C*, which goes into all that in detail.

Turtle Graphics

Turtle Graphics was first developed for the LOGO language, which was supposed to simplify programming for children. It consisted of a *turtle*, which was displayed on the graphics screen and could be pointed and moved by simple commands. A variation of Turtle Graphics has been found to be useful for generating von Koch and other fractal curves. We have global variables that tell us the direction that the turtle is pointing, its coordinates, and the size for a step of turtle movement. There are only three functions that we use, described below.

Point

The input parameters to the function point are the coordinates of the beginning and end points of a line. The function determines the turtle angle in relation to the x axis if the turtle is facing in the direction of the line defined by the input points. This angle, in degrees, is returned by the function. The function is included in Figure 3-2.

Turn

For this function, you specify an angle through which you want the turtle to turn. Positive angles are counter-clockwise and negative angles are clockwise. The function adds the specified angle to the global variable that defines the current turtle angle. The function is included in Figure 3-2.

Step

This function moves the turtle position by one step. The step length is defined by the parameter *turtle_r*. The function makes use of the current turtle position coordinates *turtle_x* and *turtle_y* and the turtle direction angle *turtle_theta* to determine the new position coordinates after the step has been taken. The function is included in Figure 3-2.

Saving and Compressing Display Data

As much as three days can be required to draw some of the fractal pictures that will be discussed in the following chapters, particularly if you do not have a math coprocessor. Since we don't want to spend another three days of drawing time whenever we want to display one of these pictures, it is essential that we have a quick, simple means of saving a picture that is on the screen to a disk file and then quickly restoring it to the screen whenever we need it. There are two instances where we need this feature:

1. When we are investigating deep within the Mandelbrot set or a similar set, and wish to start with a set that we have already generated and create an expansion of a particular part of it. The program needs to quickly display the last set generated, from a saved disk file, and then use the cursor to select the portion of it to create a new picture.

2. When we are in the process of drawing a lengthy picture and discover that we need our computer for something else, we need the capability to save the partially drawn picture to a disk file so that we can later recover it and proceed from where we left off, rather than having to begin the drawing process all over again.

Format for Saving a Screen File

The format chosen for the file that results from saving a screen is the .PCX file developed by ZSoft and is used with their PC Paintbrush and other drawing programs. This is a widely used format that permits your screens to be edited with PC Paintbrush, and programs are available from ImageSet Corp in San Francisco, CA, which can convert these files directly to slides or photographic artwork suitable for publication. ZSoft is extremely cooperative in making information on this format available to those who want to write compatible software to use it. Shannon's technical support group provided me with a pamphlet giving full technical details on the .PCX format. Functions are listed in Figure 3-2 that permit you to save a screen to a disk file using the .PCX format and also to read the file back from disk to your display. In addition there is an excellent public domain program called ZS that can be used to display any or all of your EGA .PCX files or run a slide show of them. This program is available on bulletin boards or by sending $10.00 to :

> Bob Montgomery
> 132 Parsons Rd.
> Longwood, FL 32779

The contents of the .PCX file are described in Appendix B. I have added the floating point numbers *XMax, XMin, YMax, YMin, Pval,* and *Qval* which are needed as parameters for continuation of the Mandelbrot and other sets. They are not part of the original ZSoft format, but are inserted in an empty part of the file header. Right now, files generated with the *save_screen* procedure described below are fully compatible with ZSoft programs, but this could change with a new release.

Function to Save a Screen

Data is read from the screen, horizontally from left to right, starting at the pixel position for the upper left corner. For EGA and VGA, which have multiple memory planes, a line is read of the color red (to the end of the window boundary), then the green information for the same line is read, and finally the blue. The procedure that is developed below will only work if the horizontal pixel boundaries are at a byte interface (the column number must be divisible by eight.) Data is *run-length* encoded in the following manner. If the byte is unlike the ones on either side of it, and if its two most significant bits are not *11*, it is written to the file. Otherwise a count is made of the number of like bytes (up to 63) and this count is ANDed with C0H and the result written to the file, followed by the value of the byte. If there are more that 63 successive like bytes, the count for 63 and the byte are written and then the count begins all over again. (Note that the case for a singular byte having the two most significant bits 1 is handled by writing a count of one followed by the byte value.)

The function to save an EGA mode 16 display to a disk file, *save_screen*, is included in Figure 3-2. The parameters passed to this function are the coordinates of the upper left corner of the window to be saved, the coordinates of the lower right corner, and the name of the file in which the screen data is to be stored. No protection is afforded for values that are outside the screen limits; the programmer must provide this in the calling program. Also, although any pixel location on the screen may be specified, the x value of each corner, as used by the program, is set up to be a byte boundary, which may be as much as seven pixels off from the specified value. Normally, this will not cause a problem; if it does, the programmer should assure that the x values are divisible by eight. The program assumes that the file name, which is passed to it as a parameter, consists of six letters, followed by two numbers.

The first thing to note is that all writing to the file is going to be on a byte by byte basis. In the heading, we are going to have to write out several integers and several floating point numbers. Observe how the two record types have been set up to have the same number of bytes as integers and real numbers in single format, respectively. It is then possible to use these record types to extract byte by byte information from integers or single floating point numbers to send out to the file.

The single type for real numbers was selected because it is compatible with the format and length of floating point numbers in the C language, so that displays saved using this procedure will be fully compatible with those saved and restored by C functions in my book *Fractal Programming in C*. However, the single type in Turbo Pascal requires that you be in the 8087 (real or emulated) mode when you compile; otherwise you'll get an error message.

The program first turns on a sound to indicate that the saving process is underway. It then tries to open and close the file having the designated file name in the read mode, with the error reporting function turned off. If the file can be opened (the file does indeed exist) the procedure assumes that the file already contains valuable data and therefore increments the two digit ending and tries again. The loop continues until a file name is generated that cannot be opened, indicating that the file does not exist. This file is then opened in the write mode for saving of the current screen. Note that if the two digits get to *99* without a non-existent file being found, the loop gives up. The file will then be written with the last two characters *:0*, and if this file already exists, it will be written over. When the *save_screen* terminates, it returns the two digits determined above for the end of the file name to the calling program in the form of a string, so that they are available when needed by the calling program.

The procedure then continues by writing the appropriate header information, including the palette information, which is generated as described in Appendix B. Once the header is complete, a loop begins that reiterates for every line of the display from the first one specified by *y1*, to the last one, specified by *y2*. At the beginning of this loop, the function gets the first byte of eight pixels from the first plane of the EGA screen. This is stored in *old_ch*. Next, another for loop is begun, which reads one byte at a time from the beginning to the end of the line for each of the four memory planes. After each read, action is taken based upon comparing the read character with the previous character, which was stored in *old_ch*. For the very last pass through the loop, instead of reading a byte (which wouldn't be there anyway, since we have already finished the line), we create an artificial character that is always different from *old_ch*, which forces a write out to the file of the previous data. On each pass through the loop, we check the value of the address variable *add2* (which is incremented at the end of each pass). If it is equal to the value representing the end of the line, we reset it to the starting value and also increment j, which determines which memory plane is read.

After the character is read, we check it against the previous character value; if it is the same and if *number*, which stores the number of like characters so far encountered, is less than 63, we simply increment *number* and return for the next pass through the loop. If *number* had reached 63, or if the character read differs from the previous character, we write out to the file. If number is one, indicating that the previous character is unlike those on either side of it, and if the value in *old_ch* does not have its two most significant bits equal to one, we simply write this value out to the file. If the value in *old_ch* was repeated, or if its two most significant bits are ones, we first write out the value of number with its two most significant bits set to one. We then write out the value in *old_ch*. We then reset number to one and are ready for another pass through the loop. When this loop and the display line loop have been completed, the disk file is closed and the sound turned off.

Function to Restore an EGA Screen

The function *restore_screen*, which takes a disk file saved by the procedure described above and displays it on the screen, is also listed in Figure 3-2. The function begins by attempting to open the file whose designation is passed through the parameter *file_name*. If the file does not exist, the function displays "Cannot find *file_name*," where *file_name* is the designated name, and then exits to terminate whatever program is running. If the file does exist, the first character of the header is read. If it is not the password character 0AH for .PCX files, the function displays "*file_name* is not a valid ZSoft file" and then exits to terminate the program. If the file appears to be a valid one, the function begins reading the header data.

In order to read data on integers and real numbers properly, we set up two buffers and define pointers that point to their starting addresses. Note that it is necessary to use the *New* procedure to allocate memory space for these buffers; if you fail to do this, very strange things will happen when you run the program.

The data for an integer or real number is first read from the disk file and accumulated in the appropriate buffer. Then the buffer contents are transferred to a variable with the appropriate type casting. Before the color data is read, the computer is set to EGA display mode 16 and the graphics initialized. The function then begins to read the header information. The window top left and bottom right coordinates are read and stored. The color data for each palette is read from each triple and converted to an IBM EGA format color word, which is sent to the appropriate palette by our own *setEGApalette* procedure.

The limits and P and Q values for Mandelbrot and similar sets are read. Dummy reads then take place to get to the end of the header block. The function then sends data to set up the registers of the EGA for reception of color data. Next, a *for* loop is begun for reading and displaying data for each line of the display from the top to the bottom of the window. Parameters are set up for the initial address of screen memory and for the address of the end of the current line. The parameter j is set to zero so that data will be sent to the first memory plane.

The function then begins a *while* loop that reads data from the disk file, character by character. If the character does not have its two most significant bits set to one, it is simply sent to display memory and the memory address incremented. If the first two bits are one, these are stripped and the remainder of the byte is used as a counter.

The next character is read from disk and repeatedly sent to display memory and the memory address incremented and the counter decremented each time until the counter reaches zero. After each incrementing of the memory address, the address is checked against the value for line end and if that value has been reached, the memory address is reset to the beginning of the line and the memory plane indicator is incremented. When this indicator reaches 4, all memory planes have been completed for the designated line, so the *while* loop is terminated.

When all lines have been completed, the for loop terminates, the EGA registers are reset, the disk file is closed, and the function returns with a value of 0.

Function to Display and Move the Cursor

The improved Mandelbrot set program, as well as the programs for generating similar sets for dragon and phoenix functions (which will be listed later), make use of the *move_cursor* function to position a cursor on the screen and use it to select the limits for a rectangle that defines the limits of the next screen to be generated. This function is listed in Figure 3-2.

The parameters that are passed to this function are a type number, a number for the color of the cursor, and the minimum column and row positions. Usually the *move_cursor* function is first called initially with type zero and the minimum column row and row both zero. This permits the operator to establish the top left hand corner of a desired area, whose parameters are returned through *XMin* and *YMax*. The next call to the function uses the values found above as minimums so that the lower right cursor cannot travel above the upper left corner. Exiting this cursor call sets values for *XMax* and *YMin*. There is also a third cursor mode which produces an arrow and returns the values of *Pval* and *Qval*.

The primary problem with moving the cursor is to save the original display data so that it can be restored every time the cursor moves. To do this, we establish three buffers and begin the procedure by storing the image data near the bottom of the screen (where parameter data is to be displayed) in two of them. Next the procedure enters a loop that continues until an *Enter* key is read from the keyboard. It begins by getting the image of the display at the currently desired location for the cursor and storing it in one of the buffers.

Depending upon the type value, we draw the cursor, pixel by pixel, on the display. We then read a key from the keyboard. The original display is then restored at the position where the cursor was. The procedure then checks whether the character received from the keyboard was a null. If it was, then a key having an extended keycode was activated, and the procedure reads the second byte of this code. It then checks whether one of the arrow keys was pressed, and if so, modifies the cursor position variables to move the cursor location one unit in the specified direction if this does not exceed the limits of the display.

If the key did not have an extended code, the procedure loads registers and does an interrupt to determine if the shift key was held down. If it was, and one of the arrow keys was also depressed, the values of the cursor location variables are modified to produce a step of ten units in the selected direction, if this does not exceed the boundaries of the display.

Next, the procedure writes out parameter data to the display; the parameters being selected according to the cursor type. Turbo Pascal has two methods of writing to the video screen. If the parameter *DirectVideo* is *True*, direct writes to the screen occur. This is the fastest method, and it is the default method, but unfortunately it does not work with graphics screens. The other method, writing through the ROM BIOS routines, is slower, but works for either text or graphics screens. So to display variable values at the bottom of the graphics screens that use the *move_cursor* procedure, we have to set *DirectVideo* to *False* if we want to use the normal write procedures.

The loop continues until the *Enter* key is struck. At that point the procedure terminates, with the latest values of the parameters and cursor position stored in global variables.

Using Units with Turbo Pascal

Turbo Pascal makes use of the concept of units, whereby you can put together and compile a group of related functions and/or procedures and then use them in conjunction with any program that you want to create. You begin by using the word unit, followed by the name that you want to assign to the unit. Next comes the key word interface. Following this is all of the information that you wish to make available to the outside world from your unit. This includes, first, a definition of all variables, constants, etc., which you wish to be global to your program as well as the functions and procedures within the unit. Next comes a listing of all of the procedure and function headings. These are the same as would normally exist for any procedure or function that you would include in a program. Next comes the keyword implementation, then all of the procedures and functions that you want to include in the unit. Having already defined each procedure or function in the interface section, you can use a short form of the heading for each in the implementation section, such as:

```
procedure setEGApallete;
```

You can use the full heading format, however, and it probably makes it easier to understand what's going on.

Once you have completed and debugged your unit, you can compile it by striking the F9 key. The unit automatically compiles as a .TPU type file. Figure 3-2 is a complete listing of a unit called fractal that contains all of the functions and procedures that have been described in this chapter. Using the functions and procedures from a unit in one of your programs is simple. You simply include at the beginning of the program a statement of the form:

```
uses fractal;
```

You then have access to all of the global variables, functions, and procedures that were included in the unit. Since you don't have to re-compile these functions and procedures every time you modify your program, you will save a lot of time by using strategically designed units.

Figure 3-2. Listing of Fractal Unit

```
unit Fractal;

interface

const

    PALETTE: array[0..15] of byte =
        (0,1,2,3,4,5,20,7,56,57,58,59,60,61,62,63);
var
turtle_r,turtle_x,turtle_y,turtle_theta,XMax,XMin,YMax,YMin,
    TXMax, TXMin,TYMax,TYMin,Pval,Qval: real;
CURSOR_X,CURSOR_Y: integer;

procedure setEGApalette(pal:integer; color:integer);
```

```pascal
function read_screen(address: longint; color_plane:
    integer):byte;
procedure display(address: longint; color_plane: integer; ch: byte);
function point(x1: real; y_one: real; x2: real; y2: real): real;
procedure step;
procedure turn(angle: real);
function save_screen(x1: integer; y1: integer; x2: integer;
    y2:integer; file_name: string): string;
function restore_screen(file_name:string):integer;
procedure move_cursor(C_type: integer; color: integer; min_col:
    integer; min_row: integer);

implementation
uses CRT,Graph,Dos;
procedure setEGApalette(pal:integer; color:integer);

    var
        regs: Registers;

    begin
        with regs do
        begin
            PALETTE[pal] := byte(color);
            AH := $10;
            AL := 0;
            BH := color;
            BL := pal;
            intr($10,regs);
        end;
    end;


function read_screen(address: longint; color_plane:
    integer):byte;

var
    pixel_data: byte;

begin
    Port[$3CE] := 4;
    Port[$3CF] := color_plane;
    Port[$3CE] := 5;
    Port[$3CF] := 0;
    pixel_data := byte(Mem[$A000:address]);
    read_screen := pixel_data; end;

procedure display(address: longint; color_plane: integer;
    ch: byte);

    var
        dummy: byte;
```

```
        k: byte;
        m: char;
    begin
        dummy := Mem[$A000:address];
        Port[$3C4] := 2;
        Port[$3C5] := $01 shl color_plane;
        Mem[$A000:address] := ch;
    end;


function point(x1: real; y_one: real; x2: real; y2: real): real;

    var
        theta: real;
    begin
        if (x2 - x1) = 0 then
            if y2 > y_one then
                theta := 90
            else
                theta := 270
        else
            theta := arctan((y2-y_one)/(x2-x1))*57.295779;
        if x1>x2 then
            theta := theta + 180;
        point := theta;
    end;


procedure step;

    begin
        turtle_x := turtle_x +
            turtle_r*cos(turtle_theta*0.017453292);
        turtle_y := turtle_y +
        turtle_r*sin(turtle_theta*0.017453292);
    end;

procedure turn(angle: real);

    begin
        turtle_theta := turtle_theta + angle;
    end;

function save_screen(x1: integer; y1: integer; x2: integer;
    y2:integer; file_name: string): string;

    type
        intRec = record
                    lo,hi: byte;
                end;
        floatRec = record
```

```
                        first,second,third,fourth: byte;
            end;
var
    FileExists: boolean;
    B: byte;
    GraphDriver, GraphMode,i,j,k,add1,add2,number,
        line_length, ending, start_line, end_line: integer;
    ch,ch1,old_ch,num_out,red,green,blue,color: byte;
    fsave: FILE OF byte;
    floating: single;
    file_no: string[3];

begin
    sound (256);
    repeat
        {$I-}
        Assign(fsave,file_name);
        Reset(fsave);
        Close(fsave);
        {$I+}
        FileExists := (IOResult = 0);
        if FileExists then                 begin
            inc(file_name[8]);
            if file_name[8] >= char($3A) then
            begin
                file_name[8] := char($30);
                inc(file_name[7]);
            end;
        end;
    until not FileExists or (file_name[7] > char($39));
    Assign(fsave,file_name);
    file_no := Copy(file_name,7,2);
    Rewrite(fsave);
    ch := $0A;
    Write(fsave,ch);
    ch := $05;
    Write(fsave,ch);
    ch := $01;
    Write(fsave,ch);
    ch := $04;
    Write(fsave,ch);
    B := intRec(x1).lo;
    Write(fsave,B);
    B := intRec(x1).hi;
    Write(fsave,B);
    B := intRec(y1).lo;
    Write(fsave,B);
    B := intRec(y1).hi;
    Write(fsave,B);
    B := intRec(x2).lo;
    Write(fsave,B);
```

```
B := intRec(x2).hi;
Write(fsave,B);
B := intRec(y2).lo;
Write(fsave,B);
B := intRec(y2).hi;
Write(fsave,B);
number := 640;
B := intRec(number).lo;
Write(fsave,B);
B := intRec(number).hi;
Write(fsave,B);
number := 350;
B := intRec(number).lo;
Write(fsave,B);
B := intRec(number).hi;
Write(fsave,B);
ch := 0;
for i:=0 to 15 do
begin
    red := (((PALETTE[i] and $20) shr 5) or
        ((PALETTE[i] and $04) shr 1)) * 85;
    green := (((PALETTE[i] and $10) shr 4) or
        (PALETTE[i] and $02)) * 85;
    blue := (((PALETTE[i] and $08) shr 3) or
        ((PALETTE[i] and $01) shl 1)) * 85;
    Write(fsave,red);
    Write(fsave,green);
    Write(fsave,blue);
end;
ch := $00;
Write(fsave,ch);
ch := $04;
Write(fsave,ch);
start_line := x1 div 8;
end_line := x2 div 8 + 1;
line_length := end_line - start_line;
ending := start_line + line_length * 4 + 1;
B := intRec(line_length).lo;
Write(fsave,B);
B := intRec(line_length).hi;
Write(fsave,B);
number := 1;
B := intRec(number).lo;
Write(fsave,B);
B := intRec(number).hi;
Write(fsave,B);
floating := XMax;
B := floatRec(floating).first;
Write(fsave,B);
B := floatRec(floating).second;
Write(fsave,B);
```

```
B := floatRec(floating).third;
Write(fsave,B);
B := floatRec(floating).fourth;
Write(fsave,B);
floating := XMin;
B := floatRec(floating).first;
Write(fsave,B);
B := floatRec(floating).second;
Write(fsave,B);
B := floatRec(floating).third;
Write(fsave,B);
B := floatRec(floating).fourth;
Write(fsave,B);
floating := YMax;
B := floatRec(floating).first;
Write(fsave,B);
B := floatRec(floating).second;
Write(fsave,B);
B := floatRec(floating).third;
Write(fsave,B);
B := floatRec(floating).fourth;
Write(fsave,B);
floating := YMin;
B := floatRec(floating).first;
Write(fsave,B);
B := floatRec(floating).second;
Write(fsave,B);
B := floatRec(floating).third;
Write(fsave,B);
B := floatRec(floating).fourth;
Write(fsave,B);
floating := Pval;
B := floatRec(floating).first;
Write(fsave,B);
B := floatRec(floating).second;
Write(fsave,B);
B := floatRec(floating).third;
Write(fsave,B);
B := floatRec(floating).fourth;
Write(fsave,B);
floating := Qval;
B := floatRec(floating).first;
Write(fsave,B);
B := floatRec(floating).second;
Write(fsave,B);
B := floatRec(floating).third;
Write(fsave,B);
B := floatRec(floating).fourth;
Write(fsave,B);
ch := byte(' ');
for i:=94 to 127 do
```

```
        Write(fsave,ch);
    for k:=y1 to y2 - 1 do
    begin
        add1 := 80*k;
        number := 1;
        j := 0;
        add2 := (start_line);
        old_ch := read_screen(add1 + add2,0);
        inc(add2);
        for i:=add2 to ending-1 do
        begin
            if i = ending - 1 then
                if old_ch = 0 then
                    ch:= 1
                else
                    ch:= 0
            else
            begin
                if add2 = end_line then
                begin
                    inc(j);
                    add2 := (start_line);
                end;
                ch := read_screen(add1 + add2,j);
            end;
            if ((ch = old_ch) and (number < 63))
                then
                inc(number)
            else
            begin
                num_out := byte(number or $C0);
                if ((number <> 1) or ((old_ch and
                    $C0) = $C0)) then
                Write(fsave, num_out);
                Write(fsave,old_ch);
                old_ch := ch;
                number := 1;
            end;
            inc(add2);
        end;
    end;
    Close(fsave);
    nosound;
    save_screen := file_no;
end;


function restore_screen(file_name:string):integer;

    type
        buffer = array[0..3] of byte;
```

```pascal
    buffer2 = array[0..1] of byte;
var
    bufptr: ^buffer;
    bufptr2: ^buffer2;
    ch,ch1,red,green,blue,color: byte;
    line_length,l_end: char;
    GraphDriver,GraphMode,line_end,i,j,k,m,pass,x1,
        y1,x2,y2: integer;
    fsave: FILE OF byte;
begin
    New(bufptr);
    New(bufptr2);
    {$I-}
    Assign(fsave,file_name);
    Reset(fsave);
    {$I+}
    if IOResult <> 0 then
    begin
        writeln('Can''t find ',file_name);
        Exit;
    end;
    Read(fsave,ch);
    if ch <> $0A then
    begin
        writeln(file_name,' is not a valid ZSoft file.');
        Close(fsave);
        Exit;
    end;
    for i:= 1 to 3 do
        Read(fsave,ch);
    for i:=0 to 1 do
        Read(fsave,bufptr2^[i]);
    x1 := integer(bufptr2^);
    for i:=0 to 1 do
        Read(fsave,bufptr2^[i]);
    y1 := integer(bufptr2^);
    for i:=0 to 1 do
        Read(fsave,bufptr2^[i]);
    x2 := integer(bufptr2^);
    for i:=0 to 1 do
        Read(fsave,bufptr2^[i]);
    y2 := integer(bufptr2^);
    for i:=12 to 15 do
        Read(fsave,ch);
    GraphDriver := 4;
    GraphMode := EGAHi;
    InitGraph(GraphDriver,GraphMode,'');
    for i:=0 to 15 do
    begin
        Read(fsave,ch);
        red := ch div 85;
```

```
    Read(fsave,ch);
    green := ch div 85;
    Read(fsave,ch);
    blue := ch div 85;
    color := ((red and $01) shl 5) or ((red and
        $02) shl 1) or ((green and $01) shl 4)
        or (green and $02) or ((blue and $01)
        shl 3) or ((blue and $02) shr 1);
    setEGApalette(i,color);
end;
for i:=64 to 69 do
    Read(fsave,ch);
for i:=0 to 3 do
    Read(fsave,bufptr^[i]);
XMax := single(bufptr^);
for i:=0 to 3 do
    Read(fsave,bufptr^[i]);
XMin := single(bufptr^);
for i:=0 to 3 do
    Read(fsave,bufptr^[i]);
YMax := single(bufptr^);
for i:=0 to 3 do
    Read(fsave,bufptr^[i]);
YMin := single(bufptr^);
for i:=0 to 3 do
    Read(fsave,bufptr^[i]);
Pval := single(bufptr^);
for i:=0 to 3 do
    Read(fsave,bufptr^[i]);
Qval := single(bufptr^);
for i:= 94 to 127 do
    Read(fsave,ch);
Port[$3CE] := 8;
Port[$3CF] := $FF;
Port[$3CE] := 3;
Port[$3CF] := $10;
for k:=y1 to y2-1 do
begin
    i := k*80 + (x1 div 8);
    line_end := k* 80 + (x2 div 8)+1;
    j := 0;
    while j < 4 do
    begin
        Read(fsave,ch1);
        if (ch1 and $C0) <> $C0 then
        begin
            display(i, j, ch1);
            inc(i);
            if i >= line_end then
            begin
                inc(j);
```

```
                                i := k*80 + (x1 div 8);
                        end;
                end
                else
                begin
                    ch1 := ch1 and $3F;
                    pass := ch1;
                    Read(fsave,ch);
                    for m:=0 to pass-1 do
                    begin
                        display(i, j, ch);
                        inc(i);
                        if i >= line_end then
                        begin
                            inc(j);
                            i := k*80 + (x1 div 8);
                        end;
                    end;
                end;
            end;
        end;
        Port[$3CE] := 3;
        Port[$3CF] := 0;
        Port[$3CE] := 8;
        Port[$3CF] := $FF;
        Close(fsave);
        restore_screen := x2;
    end;


procedure move_cursor(C_type: integer; color: integer;
    min_col: integer; min_row: integer);
var
    screen_buffer: array[0..512] of byte;
    i,j,temp: integer;
    limit: array[0..6] of integer;
    regs: Registers;
    ch1: char;

begin
    DirectVideo := false;
    limit[0] := 10;
    limit[1] := 8;
    limit[2] := 9;
    limit[3] := 9;
    limit[4] := 11;
    limit[5] := 13;
    limit[6] := 13;
    with regs do
        repeat
```

```
GetImage(CURSOR_X,CURSOR_Y,CURSOR_X+15,CURSOR_Y+15,
    screen_buffer);
            case C_type of
                0: begin
                    for i:=0 to 15 do
                    begin
                        PutPixel(CURSOR_X+i,CURSOR_Y,15);
                        PutPixel(CURSOR_X,CURSOR_Y+i,15);
                    end;
                    end;
                1: begin
                    for i:=0 to 15 do
                    begin
                        PutPixel(CURSOR_X+i,CURSOR_Y+15,15);
                        PutPixel(CURSOR_X+15,CURSOR_Y+i,15);
                    end;
                    end;
                2: begin
                    for j:=0 to 6 do
                    begin
                        for i:=j to limit[j] do
                        begin
                            if (i=8) and (j=5) then
                                i := 10;
                            if (i=8) and (j=6) then
                                i := 12;
                            PutPixel(CURSOR_X+j,CURSOR_Y+i, 15);
                        end;
                    end;
                    end;
                end;              ch1 := ReadKey;
        if ch1 <> char($0D) then
        begin
            PutImage(CURSOR_X,CURSOR_Y,screen_buffer,0);
            AH := 2;
            intr($16,regs);
            if ch1 = char($00) then
            begin
                ch1 := ReadKey;
                case ch1 of
                 'M': begin
                        if CURSOR_X < 639 then
                            inc(CURSOR_X);
                    end;
                 'K': begin
                        if CURSOR_X > min_col then
                            dec(CURSOR_X);
                    end;
                 'H': begin;
                        if CURSOR_Y > min_row then
                            dec(CURSOR_Y);
```

```
                end;
        'P': begin
                if CURSOR_Y < 335 then
                    inc(CURSOR_Y);
              end;
          end;
    end
    else
    begin
        AH := 2;
        intr($16,regs);
        if (AL and $03) <> 0 then
        begin
            case ch1 of
                '8':  begin
                        if CURSOR_Y > min_row
                            then
                            CURSOR_Y := CURSOR_Y
                                - 10;
                      end;
                '4':
                    begin
                        if CURSOR_X > min_col
                            then
                            CURSOR_X := CURSOR_X
                                - 10;
                    end;
                '6':
                    begin
                        if CURSOR_X < 613 then
                            CURSOR_X := CURSOR_X
                                + 10;
                    end;
                '2':
                    begin
                        if CURSOR_Y < 323 then
                            CURSOR_Y := CURSOR_Y
                                + 10;
                    end;
            end;
        end;
    end;
    case C_type of
        0: begin
            TXMin := XMin + (XMax - XMin)/
                639*(CURSOR_X);
            TYMax := YMax - (YMax - YMin)/
                349*CURSOR_Y;
            GotoXY(5,24);
            write('XMin: ',TXMin:7:6,'  YMax:
                ',TYMax:7:6,'     ');
```

```
                            end;
                  1: begin
                       TXMax := XMin + (XMax - XMin)/
                           639*(CURSOR_X + 16);
                       TYMin := YMax - (YMax - YMin)/
                           349*(CURSOR_Y + 16);
                       GotoXY(41,24);
                       write('XMax: ',TXMax:7:6,'  YMin:
                           ',TYMin:7:6,'     ');
                     end;
                  2: begin
                       Pval := XMin + (XMax - XMin)/639*
                           CURSOR_X;
                       Qval := YMax - (YMax - YMin)/
                           349*CURSOR_Y;
                       GotoXY(5,24);
                       write('Pval: ',Pval:7:6,'   Qval:
                           ',Qval:7:6);
                     end;
                end;
             end;
          until ch1 = char($0D);
       end;
end.
```

Observing the Parameters in a Display File

The values of some or all of *XMax, YMax, XMin, YMin, P,* and *Q* are essential in proceeding from one of the Mandelbrot set or similar figures to another more expanded one, in completing a partially generated figure that has been saved on disk, and in generating one of the Julia or similar sets from the appropriate map figure. We also would like to know at times the actual color values used in generating a figure. All of this information is stored on the disk file which stores the figure for future use. The program *params* asks for a figure file name and then displays that file on the screen. It then overwrites on it the parameters given above and all of the color palette values.

The figure is not displayed primarily for use at this time, but simply to give you an opportunity to assure that you asked for the right file name. Thus it doesn't matter much if it gets partially covered up. By then, you have already identified the figure, and in any case, after *params* is done there is usually enough of the figure left displayed for satisfactory identification. Figure 3-3 lists the params program.

Figure 3-3. Function to Show Figure Parameters

```
program params;

uses CRT,Graph,Fractal;
var
    file_name: string[20];
    column,graphDriver,GraphMode,color,i,palette_register:
integer;
    ch1: char;
begin
    DirectVideo := false;
    writeln;
    write('Enter file name: ');
    readln(file_name);
    column := restore_screen(file_name);
    if column = 0 then
        exit;
    for i:=0 to 15 do
            writeln('Palette #',i,' = ',PALETTE[i]);
    writeln('XMax = ',XMax:7:6);
    writeln('XMin = ',XMin:7:6);
    writeln('YMax = ',YMax:7:6);
    writeln('YMin = ',YMin:7:6);
    writeln('P = ',Pval:7:6);
    writeln('Q = ',Qval:7:6);
    ch1 := ReadKey;
end.
```

Selecting Colors

Sometimes, the best laid plans for creating beautiful colors go astray, and the resulting figure is horribly different from what you antici-pated. Of course, you could go back to the original program, change the *setEGApalette* statements or in some other way modify the way in which you specify that colors be generated. The program described in this section provides an easier method. It will display a selected screen file on the screen and allow you to change each of the sixteen palettes to any of the 64 shades available with the EGA. When you are done, it will save the display together with the new color designations in a new disk file. The program *colors* is listed in Figure 3-4.

At the beginning, the program asks you for a file name. It can read in any *.pcx* file on which you have stored a display. It then permits you to change all of the display colors. Once you start changing colors with the left and right cursor arrows, a legend will appear at the bottom of the screen giving the current palette number and color number. Don't be dismayed; when you are finished changing colors, the display will be redrawn so that the legend will not appear in your new display file. You begin the process by entering a palette number (between 0 and 15), followed by hitting the *Ent* key. The color number will be automatically set to that of the current color shade for the selected palette.

You can change the color number by hitting the right or left arrow keys. Each time you hit the right arrow, the color number will increase by one and the color of the selected palette on the display will change accordingly. Each time you hit the left arrow, the color number will decrease by one and the color of the selected palette on the display will change accordingly. If you hold down an arrow key to scan through color changes and you go too far, you can reverse direction by using the other arrow key. Have no fear, you cannot get out of the permissible color range of 0 to 63; if you go beyond 63, the color number returns to 0, and if you go below zero, the color number returns to 63.

Once you find the color shade that you like, hitting *Ent* freezes that color into the selected palette. You are then ready to enter another palette number. When you have the picture colored exactly as you want it, entering a palette number greater than 15 terminates the program.

The screen is rewritten to get rid of the legend at the bottom. The rewritten screen appears in the original colors, but your color modifications are saved and will permanently become part of the new file.

The new display will be saved in a file called *colorsnn.pcx,* where *nn* is a pair of digits from 00 to 99. The program will automatically start out with 00 and search sequentially for a pair of digits that you have not used yet. When it finds them, they will be used for the file that is currently being saved. The colors program is listed in Figure 3-4.

Figure 3-4. Function to Change Display Colors

```
program colors;

uses CRT,Graph,Fractal;

const     file_name2: string[20] = ('colors00.pcx');
var
    file_name: string[20];
    column,graphDriver,GraphMode,color,i,palette_register:
        integer;
    chl: char;
    SAVER: array[0..15] of byte;
    dummy: string[3];

begin
    DirectVideo := false;
    writeln;
    write('Enter file name: ');
    readln(file_name);
    column := restore_screen(file_name);
    if column = 0 then
        exit;
    for i:=0 to 15 do
        SAVER[i] := PALETTE[i];
    repeat
        GotoXY(10,23);
        write('Palette:     ');
        GotoXY(19,23);
        readln(palette_register);
        if palette_register < 16 then
        begin
            color := PALETTE[palette_register];
            repeat
                chl := ReadKey;
                if chl = char(0) then
                begin
                    chl := ReadKey;
                    if chl = char(77) then
                        inc(color);
                    if chl = char(75) then
                        dec(color);
```

```
              if color > 63 then
                  color := 0;
              if color < 0 then
                  color := 63;
              GotoXY(10,24);
              writeln('Color: ',color,'   ');
              setEGApalette(palette_register,color);
          end;
      until ch1 = char($0D);
    end;
    saver[palette_register] := color;
  until palette_register > 15;
  column := restore_screen(file_name);
  for i:=0 to 15 do
      PALETTE[i] := SAVER[i];
  dummy := save_screen(0,0,639,349,file_name2);
end.
```

4

The Lorenz and Other Strange Attractors

In 1962, Edward Lorenz was attempting to develop a model of the weather when he observed some strange discrepancies in the behavior of his model. When he attempted to restart the model at a point part way through the original computer run, the results, although apparently starting at the same point, diverged farther and farther from the original run as time went on. He verified that this was not a computer error, but rather was caused by the fact that he had reentered the data to only three decimal place accuracy, whereas the computer data at that point in the original computer run was saved to six decimal places. Lorenz simplified down to a model consisting of three differential equations, which also describe the flow of fluid in a layer of fluid having a uniform depth and a constant temperature difference between the upper and lower surfaces. The equations are:

```
dx/dt = 10(y - x)       (Equation 4-1)

dy/dt = xz + 28x -y     (Equation 4-2)

dz/dt = xy - (8/3)z     (Equation 4-3)
```

When Lorenz laboriously calculated a number of values for these equations on a primitive computer, he discovered the first of the strange attractors, and created the foundation for the discipline of

"Chaos," which is creating drastic changes in all fields of science, and of which the principle drawing tools are fractals.

Strange Attractors

What is a strange attractor? To answer this question, we must first plot a candidate set of equations in phase space, a space of enough dimensions to permit representing each solution of the equation set at a given time as a single point. For the Lorenz equations given above, a three dimensional phase space is needed. If the solution to this set of equations was constant throughout time, it would converge in phase space to a single point, the attractor, no matter what the initial conditions had been. If the solution converged to a periodic function which repeated over and over after fixed interval of time, the result in phase space would be some form of closed curve, the periodic attractor or limit cycle. If neither of these cases is true (yet the equation has a fully determined path through phase space, which never recurs), the resulting curve is called a strange attractor. No matter what initial conditions are specified, the solution always converges quickly to a point on this curve and continues to follow the path of the curve from there on.

The Lorenz Attractor

It's time to take a close look at the Lorenz attractor. Plate 1 shows the Lorenz attractor projected upon the YZ plane, and Plate 2 shows a three-dimensional projection. Note, however, that without most of the traditional cues that help our senses to convert a two-dimensional drawing to three dimensions, it is not too easy to understand the exact dimensional qualities of the Lorenz attractor, no matter what kind of projection we use. The color scheme used here is that the color is changed each time the value of the x coordinate crosses zero. The curves represent 8000 iterations of the equation with a time step of 0.01. Unfortunately, already the resolution of the display screen has proven inadequate to the task of separating out adjacent portions of the curve. However, no matter how good the resolution of your display, the curves will exceed its resolution capability if enough iterations are run.

These curves are a sort of encapsulation of what this new science of chaos is all about, both in its good and its bad features. You need to watch the curve being drawn and understand that although the curve seems to intersect with itself in the projections, in actual three-dimensional space it never does touch itself. The good aspect of chaos is that this simple set of equations can completely describe a very rich and complex, nonperiodic behavior. Prior to investigating this kind of equation system with modern high-speed computers, scientists postulated that such complex behavior must be the result of very complex systems of equations containing many parameters and variables, with possibly a number of random variables thrown in. Now it is known that complex behavior may often be represented in a very simple manner.

The bad aspect can be discovered in the following manner. Select a starting point somewhere on the very crowded part of the curve. Attempt to trace the path from there on. We already pointed out that the display has inadequate resolution, so that a couple of different portions of the curve double up at the most crowded places. Thus you can't be sure that you are tracing the right path, since as the adjacent curves begin to diverge, your selected path will break in two and you can't be sure which path to follow. How does this apply in the real case? There is no overlapping on the infinite resolution display, each set of initial values determines one and only one path to be followed. But there are an infinite number of paths in the vicinity of the starting point you selected, and which one will be followed depends upon how precisely you specified your initial coordinates. If you selected $x = 3.15678$, for example, you would travel a totally different path than if you had selected $x = 3.15679$. And you must remember that $x = 3.15678$ is actually $x = 3.15678000...$, so that by adding another decimal place with a value other than zero, you can always diverge to a different path altogether. This means that no matter how accurately you select the initial coordinates, if they are at all different from the real values that might exist for a natural phenomena, the value that you predict will diverge farther and farther from the "real" value as time progresses.

This is bad news for those who wish to measure some initial conditions and use them to predict long-term results. Note that measuring more precisely, so as to come closer to the correct long-term values does not work because the amount of divergence is not a function of the size of the error, but can differ widely and unpredictably.

Runge Kutta Integration

In order to solve the system of differential equations given above, we must use some numerical technique that comes up with an accurate value for x, y, and z at each time step during integration. We have chosen a time step of 0.01. Lorenz, in his original paper, used a double approximation integration technique. However, with more sophisticated computers at our disposal, we can use a more complicated integration method to produce greater accuracy. The method that will be used is the fourth-order Runge Kutta technique. This method is a one-step procedure that uses only first-order derivatives to achieve the same accuracy obtainable with an equivalent order Taylor expansion using higher-order derivatives. There are many different sets of coefficients that can be used with the Runge Kutta integration method; the coefficients that we have selected were chosen to minimize the computer time required for each iteration. Given a differential equation:

$$dy/dt = f(t,y) \hspace{2cm} \text{(Equation 4-4)}$$

once the initial condition is established, at each time step, we have

$$y_{n+1} = y_n + k_0/6 + k_1/3 + k_2/3 + k_3/6 \hspace{1cm} \text{(Equation 4-5)}$$

where

$$k_0 = h\ f(t_n,\ y_n) \qquad\qquad \text{(Equation 4-6)}$$

$$k_1 = h\ f(t_n + h/2,\ y_n + k_0/2) \qquad\qquad \text{(Equation 4-7)}$$

$$k_2 = h\ f(t_n + h/2,\ y_n + k_1/2) \qquad\qquad \text{(Equation 4-8)}$$

$$k_3 = h\ f(t_n + h,\ y_n + k_2) \qquad\qquad \text{(Equation 4-9)}$$

and h is the time step (0.01).

You will find this integration technique in the middle of the Lorenz attractor program. Note that in determining each k, the equation has to be solved for the appropriate values of t and y.

Programming the Lorenz Attractor

Figure 4-1 lists the program to generate the Lorenz attractor. Since this is our first Turbo Pascal fractal program, let's take a close look at what's involved. After the program-defining statement, a statement shows that the program uses the graphics routines CRT and Graph. The constants and variables are then defined. Next, the function degrees_to_radians is listed. This function first takes any number of degrees that was input and eliminates multiple rotations so that the result is between -360 and +360 degrees. It then applies the proper conversion to yield the equivalent number of radians. The program then begins by specifying the projection angles for the three-dimensional display, converting them to radians, and then computing the sine and cosine of each. The program doesn't allow for user changing of these angles, but it is simple to change them if you prefer another orientation, or rewrite the program so that the angles can be inserted at run time by the user. Three loops are made through the program; one to plot the projection on the YZ plane, one to plot the projection on the XY plane, and one to provide the three-dimensional projection.

The program next specifies an initial color and then defines graphics parameters and initiates the graphics mode. The program is set up to force the high-resolution EGA mode, which is a good one to use if you have an EGA or VGA. If you have a different graphics adapter or want to use another mode, you need to change the *GraphMode* parameter accordingly. You also need to change the constants *maxcol* and *maxrow* to correspond to the parameters of the mode you are using. Since the initial pixel row and column locations are 0, the value of *maxrow* should be one less than the total number of rows in your display and the value of *maxcol* should be one less than the number of columns.

Next, the line style is set to draw a solid line three pixels wide and the color set for white. After initializing x, y, and z, the program computes initial row and column values for the loop that it is in, draws the axes, and labels them. The line width is then set to one pixel wide, and the program then loops through 8000 Runge Kutta integration iterations. At the end of each iteration, new column and row locations are determined appropriately for the loop that is currently being run and a line is drawn from the previous column and row locations to the new ones. As mentioned above, the color scheme used is to change the color each time that the curve crosses the x axis. A set of tests determines whether the ends of the line segment being drawn are on opposite sides of the x axis, and if they are, increments the color. If you want to be careful, you can limit the color to the maximum value that is available for the graphics mode you are using; if you don't, the *SetColor* procedure should handle this for you. Finally, the latest column and row values are stored as the old ones and the program is ready to perform another integration iteration.

Figure 4-1. Program to Generate Lorenz Attractors

```
program lorenz;

uses CRT,Graph;

const
    maxcol = 639;
    maxrow = 349;
    rad_per_degree=0.0174533;
    third=0.333333333;
var
```

```
      x,y,z,d0_x,d0_y,d0_z,d1_x,d1_y,d1_z,d2_x,d2_y,d2_z,
      d3_x,d3_y,d3_z,xt,yt,zt,dt,dt2,x_angle,y_angle,z_angle,
      sx,sy,sz,cx,cy,cz,temp_x,temp_y,old_y: real;
      dummy,GraphDriver,GraphMode,i, j, row, color,col, old_row,
      old_col: integer;
      ch: char;

function degrees_to_radians(degrees:real):real;

      begin
          while degrees >= 360 do
              degrees := degrees - 360;
          while degrees < 0 do
              degrees := degrees + 360;
          degrees_to_radians := rad_per_degree*degrees;
      end;


begin

      x_angle := 45;
      y_angle := 0;
      z_angle := 90;
      x_angle := degrees_to_radians(x_angle);
      sx := sin(x_angle);
      cx := cos(x_angle);
      y_angle := degrees_to_radians(y_angle);
      sy := sin(y_angle);
      cy := cos(y_angle);
      z_angle := degrees_to_radians(z_angle);
      sz := sin(z_angle);
      cz := cos(z_angle);
      for j:=0 to 2 do
      begin
          color := 4;
          GraphDriver := 4;
          GraphMode := EGAHi;
          InitGraph(GraphDriver,GraphMode,'');
          SetLineStyle(0,$FFFF,3);
          SetColor(15);
          x := 0;
          y := 1;
          z := 0;
          if j = 0 then
          begin
              old_col := Round(y*9+320);
              old_row := Round(350 - 6.56*z);
              Line(0,348,639,348);
              Line(320,2,320,348);
              OutTextXY(628,330,'Y');
              OutTextXY(330,12,'Z');
```

```pascal
end;
if j = 1 then
begin
     old_col := Round(y*10+320);
     old_row := Round(175 - 7.29*x);
     Line(0,175,639,175);
     Line(320,2,320,348);
     OutTextXY(628,160,'Y');
     OutTextXY(330,12,'X');
end;
if j = 2 then
begin
     old_col := Round(y*9);
     old_row := Round(350 - 6.56*z);
     Line(0,348,638,348);
     Line(320,2,320,348);
     Line(320,348,648,140);
     OutTextXY(628,330,'Y');
     OutTextXY(330,12,'Z');
     OutTextXY(628,112,'X');
end;
SetLineStyle(0,$FFFF,1);
dt := 0.01;
dt2 := dt/2;
for i:=0 to 8000 do
begin
     d0_x := 10*(y-x)*dt2;
     d0_y := (-x*z + 28*x - y)*dt2;
     d0_z  := (x*y - 8*z/3)*dt2;
     xt := x + d0_x;
     yt := y + d0_y;
     zt := z + d0_z;
     d1_x := (10*(yt-xt))*dt2;
     d1_y := (-xt*zt + 28*xt - yt)*dt2;
     d1_z  :=(xt*yt - 8*zt/3)*dt2;
     xt := x + d1_x;
     yt := y + d1_y;
     zt := z + d1_z;
     d2_x := (10*(yt-xt))*dt;
     d2_y := (-xt*zt + 28*xt - yt)*dt;
     d2_z  :=(xt*yt - 8*zt/3)*dt;
     xt := x + d2_x;
     yt := y + d2_y;
     zt := z + d2_z;
     d3_x := (10*(yt - xt))*dt2;
     d3_y := (-xt*zt + 28*xt - yt)*dt2;
     d3_z := (xt*yt - 8*zt/3)*dt2;
     old_y := y;
     x := x + (d0_x + d1_x + d1_x + d2_x + d3_x) * third;
     y := y + (d0_y + d1_y + d1_y + d2_y + d3_y) * third;
     z := z + (d0_z + d1_z + d1_z + d2_z + d3_z) * third;
```

```
        if j = 0 then
        begin
            col := Round(y*9+320);
            row := Round(350 - 6.56*z);
            if col < 320 then
                if old_col >= 320 then
                    inc(color);
            if col > 320 then
                if old_col <= 320 then
                    inc (color);
        end;
        if j = 1 then
        begin
            col := Round(y*10.0+320);
            row := Round(175-7.29*x);
            if col < 320 then
                if old_col >= 320 then
                    inc(color);
            if col > 320 then
                if old_col <= 320 then
                    inc (color);
        end;
        if j = 2 then
        begin
            temp_x := x*cx + y*cy + z*cz;
            temp_y := x*sx + y*sy + z*sz;
            col := Round(temp_x*8 + 320);
            row := Round(350 - temp_y*5);
            if col < 320 then
                if old_col >= 320 then
                    inc(color);
            if col > 320 then
                if old_col <= 320 then
                    inc (color);
        end;

        SetColor(color);
        Line(old_col,old_row,col,row);
        old_row := row;
        old_col := col;
      end;
      ch := ReadKey;
    end;
end.
```

At the end of each loop, a call to *ReadKey* causes the program to wait until you hit a key before it erases the screen and begins to generate the next display. You may want to try some other color technique. For example, you could use different colors to represent the position of the curve in Figure 4-1. You may also want to insert different angles to change the viewing direction of the three-dimensional projection, but you may also have to do some additional modification of the dimensioning to keep part of the curve from falling off the edge of the display. The parameters that are used to multiply the original variable values to obtain the row and column values were determined experimentally to assure that the entire display appeared on the screen. There is nothing sacred about these, so you can change them to assure that a different projection does not go off the screen or to blow up a small section of the display for closer examination.

Another thing that you might like to investigate is the number of iterations of the inner loop. You can reduce or increase it and obtain different amounts of detail in the displays. Finally, just before the *drawLine* function, you can insert a pair of *if* statements similar to this:

```
if i >= 1400 then
if i <= 1900 then
```

These statements will cause only the iterations between 1400 and 1900 to be displayed. This is the section of the curve that Lorenz used to illustrate his original paper. You can, if you wish to speed up the program, use an if statement like:

```
if i >= 1400 then
```

and change the upper limit of the for loop to 1900 to achieve the same result. You cannot change the starting value of the *for* loop to 1400, since you will not then know what the initial values are for x, y, and z.

Other Strange Attractors

The Lorenz attractor proceeds in an orderly fashion from one point to the next as time increases, so that we can draw a good picture of it by drawing lines that connect each pair of adjacent points. Now let's consider a different kind of strange attractor. This one is a dynamical system first reported by Clifford A. Pickover. It consists of the system of equations:

$$x_{n+1} = \sin(ay_n) - z_n\cos(bx_n) \qquad \text{(Equation 4-10)}$$

$$y_{n+1} = z_n\sin(cx_n) - \cos(dy_n) \qquad \text{(Equation 4-11)}$$

$$z_{n+1} = \sin(x_n) \qquad \text{(Equation 4-12)}$$

There is no time step here. Moreover, the point in phase space described by the equations jumps about in what appears to be a totally random fashion. However, when the points for a large number of iterations are plotted, it becomes evident that there is a finite set of positions which the point described by the function can occupy, and that the point ultimately goes to this attractor irrespective of the initial conditions. Figure 4-2 lists a program for generating this strange attractor for a specific set of the parameters a, b, c, d, and e and displaying it projected on first the XY and then the YZ planes. The resulting displays are shown in Figures 4-3 and 4-4, respectively.

The pictures in Figures 4-3 and 4-4 use a modification of the program which displays points in white only. The program listing reads the color of each pixel, and goes through the bright colors as each additional hit occurs on a particular pixel. Thus, the final color of a pixel depends upon the number of times it has been hit during drawing the figure. The color changes stop with white, so white spots have been hit the most. You may be able to think of some other interesting way to color this display.

Figure 4-2. Program to Generate a Strange Attractor

```pascal
program strange;

uses CRT,Graph;

const
    max_col = 639;
    max_row = 349;

var

Xmax,Xmin,Ymax,Ymin,X,Y,Z,deltaX,deltaY,Xtemp,Ytemp,Ztemp,a,b,c,d,e: real;
    GraphDriver,GraphMode,col,row,j,color: integer;
    max_iterations,i: longint;
    ch: char;

begin
    max_iterations := 50000;
    Xmax := 2.8;
    Xmin := -2.8;
    Ymax := 2.0;
    Ymin := -2.0;
    X:= 0;
    Y:= 0;
    a := 2.24;
    b := 0.43;
    c := -0.65;
    d := -2.43;
    e := 1.0;
    GraphDriver := 4;
    GraphMode := EGAHi;
    InitGraph(GraphDriver,GraphMode,'');
    deltaX := max_col/(Xmax - Xmin);
    deltaY := max_row/(Ymax - Ymin);
    for j:=0 to 1 do
    begin
        for i:=0 to max_iterations do
        begin
            Xtemp := sin(a*Y) - Z*cos(b*X);
            Ytemp := Z*sin(c*X) - cos(d*Y);
            Z := e*sin(X);
            X := Xtemp;
            Y := Ytemp;
```

```
        if j=0 then
        begin
            col := Round((X - Xmin)*deltaX);
            row := Round((Y - Ymin)*deltaY);
        end;
        else
        begin
            col := Round((Y - Xmin)*deltaX);
            row := Round((Z - Ymin)*deltaY);
        end;
    if col>0 then
        if col<=max_col then
            if row>0 then
                if row<=max_row then
                begin
                    color := GetPixel(col,row);
                    if color = 0 then
                        color := 8;
                    if color < 15 then
                        color := color+1;
                    PutPixel(col,row,color);
                end;
        end;
        ch := ReadKey;
        ClearDevice;
    end;
end.
```

Figure 4-3. Strange Attractor Projected on XY Plane

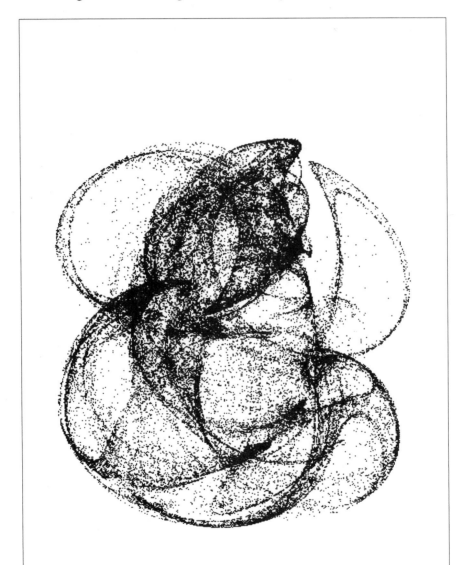

Figure 4-4. Strange Attractor Projected on YZ Plane

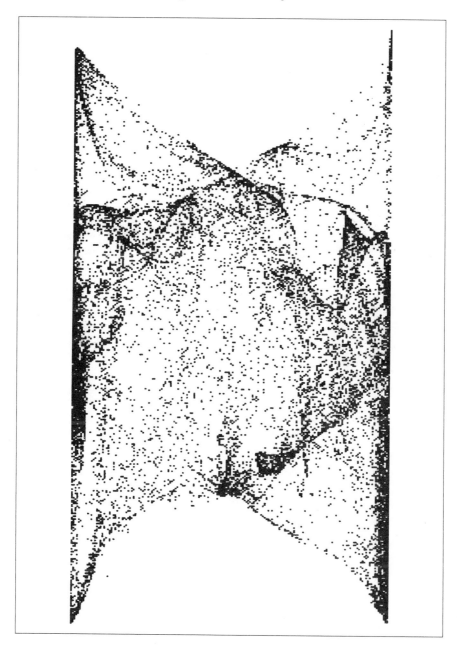

5

The Population Equation and Bifurcation Diagrams

It was in 1798 that Thomas Malthus made the first widely-known attempt to apply mathematics to the growth and decline of populations. In his paper, "An Essay on the Principle of Population as it Affects the Future Improvement of Society," Malthus presented the proposition that population, if unchecked, grows in a geometric manner while the growth of available food supplies is arithmetic. Consequently, unless strict birth control measures were introduced, Malthus foresaw extended calamity and widespread starvation. Fortunately for us, and unfortunately for the validity of Malthus' theory, improvements in food production techniques kept pace with population growth and the disaster never occurred. Consequently, Malthus' theory has been out of favor for a number of years. Just a few years ago, however, the Club of Rome commissioned the development of a massive computer model which projects population increases that will reach the limit of earthly resources and cause, by the year 2000, the kind of catastrophes which Malthus predicted. Whether one accepts these results at face value or not, the Malthusian predictions have certainly gained a new lease on life.

The Population Equation

By the early 1950s, a simplified equation for population growth was being regularly used by ecologists. The equation is:

$$x_{n+1} = rx_n(1 - x_n) \qquad \text{(Equation 5-1)}$$

Rather than simply allow the population to grow at an uncontrolled rate, the use of the $(1 - x)$ factor implies that the larger the population becomes, the more forces are applied to reduce growth. Generally speaking, using this equation (particularly if the parameter r is less than one) causes the population to reach a maximum when x is equal to 0.5. If the population dies out (x decreases to zero), the species is extinct. On the other hand, the population will also die out if such tremendous overgrowth occurs that the value of x reaches one. Strangely enough, everyone assumed that this equation was well behaved, and for a long time, no one discovered the chaotic behavior that could occur when r took on larger values. This is one of those things that common sense makes obvious, once the facts are discovered. We have things like the seven-year locusts, which have a tremendous population explosion every seven years. Surely such examples should have made us suspect that a population value could achieve a stability with more than one stable value and shift back and forth between these values in successive iterations. But it was not until 1971 that Robert Mays, at the Institute for Advanced Study at Princeton, studied this equation in detail for a wide range of values of r and at last began to come to an understanding of the complicated behavior that was hidden in the simple expression.

Bifurcation Diagrams

The best way to make sense of the really complicated behavior of the simple equation given above is through the use of a graph. These graphs are usually referred to as bifurcation diagrams. What we are going to do is travel through a range of values of r, sampling at intervals close enough so that we won't miss anything. For each r, we will start with the nominal value of 0.5 for x and do 256 iterations. After 64 iterations, x should have settled down to its final steady state conditions. We then plot the values of x associated with this r from 64 to 256 iterations. For the smaller values of r, where everything is well-behaved, we find that x has settled to a single value. But at some point, we find that there are two final values for x, then 4, then 8 and so forth. Figure 5-1 is the listing for the program to generate our bifurcation diagrams for the population equation. We loop through two iterations; the first steps r in steps of 0.005, beginning at 0.95. The resulting diagram is shown in Figure 5-2. The second pass through the loop does an expansion in the area of period three. It begins with an r of 3.55 and steps in 0.0005 increments. The resulting diagram appears in Figure 5-3.

Now let's take a close look at these diagrams. Beginning at an r value of about one, the system settles down to a single value greater than zero, which remains the same, no matter how many additional iterations we perform. This stable value increases as r is increased until, at an r in the neighborhood of three, there is a split into two stable values which alternate with each iteration. Next, there is another split and there are now four stable values between which the iterations cycle. The four then become eight, then sixteen, and so forth, until finally we reach a state of chaos in which there are so many values that we never see the repetition with the tools that we have. Although this is normally called the chaotic region, one must be careful to remember that the period doubling scenario that we have seen thus far may be continuing. There may be a repetition, but since there are 2 to a very high power different values before the cycle repeats, we are unlikely to find it.

If you like scanning tables of numbers, you can develop a program to print out a large number of values of this function and scan them to see if you can find any traces of order in the chaos. Interestingly enough, as you can see from Figure 5-3, there are windows in this chaotic behavior. At one of these, the function reverts to cycling between three stable states, then bifurcates to six, then to twelve, and so forth.

Figure 5-1. Program to Plot Bifurcation Diagrams

```
program bifurc;

uses CRT,Graph;

var

    ch: char;
    r,x,delta_r: real;
    GraphDriver, GraphMode,i,j: integer;
    row, col: longint;
begin
    r := 0.95;
    GraphDriver := 4;
    GraphMode := EGAHi;
    InitGraph(GraphDriver,GraphMode,'');
    SetColor(15);
    for j:=0 to 1 do
    begin
        delta_r := 0.005;
        if j = 1 then
        begin                   r := 3.55;
            delta_r := 0.0005;
        end;
        for col:=0 to 639 do
        begin
            x := 0.5;
            r := r + delta_r;
            i:=0;
            repeat
                x := r*x*(1-x);
{   COMPUTATION FOR rx(1-x)
                row := 349 - Round(x*350); }
{   COMPUTATION FOR x(1-x) }
                row := Round(349 -((x/r)*700));

                if i>64 then
                    if row<350 then
                        if row>=0 then
                            if col>=0 then
                                if col<640 then
                                    PutPixel(col,row,15);
                Inc (i);
            until (x>1000) or (x<-1000) or (i>255);
    end;
    ch := ReadKey;
        ClearDevice;
     end;
end.
```

Figure 5-2. Bifurcation Diagram for Population Equation

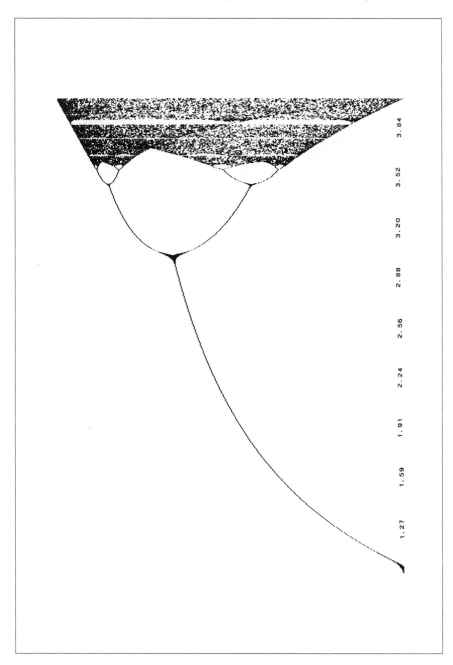

Figure 5-3. Expansion of the Bifurcation Diagram

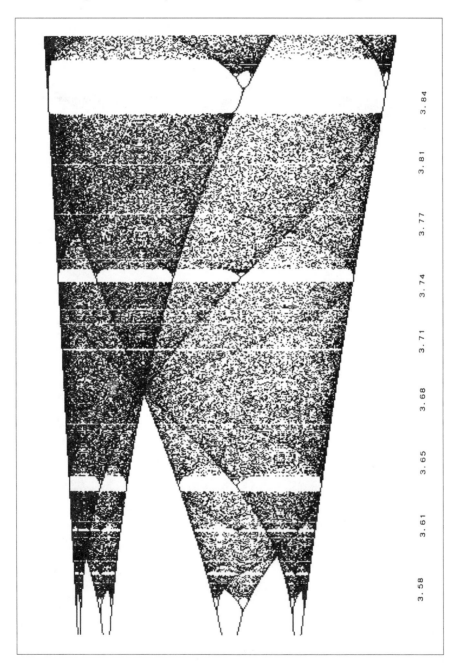

"Period Three Implies Chaos"

Robert Mays' friend James Yorke did a rigorous mathematical analysis of the behavior of the population equation and, in December 1975, together with Tien-Yien Li, published a paper called, "Period Three Implies Chaos." What Yorke and Li were able to show is that, if a function similar to the population equation has a period of three, then it has periods of every other number, n. Thus, it is rigorously established that there is an infinitely rich spectrum of results for this type of equation.

The Feigenbaum Number

If Mitchell Feigenbaum had known of the work of Robert Mays and James Yorke, or if he had been able to view Mays' bifurcation diagrams, he might never have made his significant discovery. But, in 1976, Feigenbaum was looking at the population equation from a different point of view. Consider for a moment just part of the equation:

```
y = x(1 - x)                    (Equation 5-2)
```

This equation has a maximum at $x = 0.5$. In fact, for the original population equation, in each set of bifurcations there is a value of r at which y of equation 5-2 achieves its maximum of 0.5. If Feigenbaum had been into bifurcation diagrams, he could have produced something like Figure 5-4, or the expanded version of Figure 5-5, where for each value of r we plot $x(1 - x)$ instead of $rx(1 - x)$. In these figures it becomes very clear that each set of bifurcations has one or more values of r at which y achieves its maximum. Feigenbaum was trying to determine the values of the r at which the maximum is reached for each set of bifurcations. If we start with the maximum value of x and perform 2^n iterations, where n is the number of bifurcations, we should cycle through all of the bifurcated values of x and be back to the maximum again. Thus, we have the general expression:

$$x_{max} = (r_n f)^{2m^n} (x_{max}) \qquad \text{(Equation 5-3)}$$

which, for the population equation is:

$$x_{max} = [r_n x_{max}(1 - x_{max})]^{2m^n} \qquad \text{(Equation 5-4)}$$

This equation can be solved easily for x_0, and with more difficulty for x_1, but it very quickly gets so complicated, and has so many roots, that solution becomes nearly impossible. The best method of solution is to start with a value for x below the first known root and increase it very slowly until the next root is encountered, and so forth. Feigenbaum was doing this tediously on a programmable calculator.

In looking for a way to reduce the number of calculations, he discovered a remarkable universal relationship between adjacent roots. This is expressed as the Feigenbaum number,

$$\frac{r_n - r_{n-1}}{r_{n+1} - r_n} = 4.6692016091029\ldots \qquad \text{(Equation 5-5)}$$

To be mathematically precise, the Feigenbaum number is the value which this constant reaches as n approaches infinity, so that the first few values especially differ by a considerable error. Feigenbaum later discovered that this universal ratio applies to every kind of iterated function which is characterized by having a single differentiable maximum. This includes many different kinds of algebraic equations as well as trigonometric equations, which have only a single hump of the curve.

Figure 5-6 lists a program to generate the first 17 r_n values for the population equation. It makes use of the maximum precision available from the IBM PC using Turbo Pascal. Unfortunately, you will note that the precision of the Feigenbaum number reaches its peak at around the 15th and 16th values, and that after that the computer does not have enough precision to give highly accurate values.

Figure 5-7 is a table of the values obtained by the program of Figure 5-6. However, don't expect to get these same exact values when you run the program. Because the computation is so dependent upon the exact nature of the computer hardware for its precision, your machine may give quite different results in the small decimal places. The program starts with a value of x and a step size. The equation is solved with the proper number of iterations to return to the maximized root, and then the result is compared with the value of x_{max}. Initially, x is approaching x_{max} from below, so that if the value hasn't been reached yet, we add the step and try again. We keep looping and doing this until we do exceed x_{max} and then we subtract one step value, reduce the step size, and try again. We keep on with this looping until the step is less than the minimum value that we can handle with the computer. The next root is approached decreasing from above, so that we have to reverse our test against x_{max} and reduce step size when we get below instead of above.

Finally, as we proceed to higher roots, we need to be careful to assure that the initial step size is not so large that we jump right over several adjacent roots, as the roots are getting closer together all the time. Consequently, we begin reducing the initial step size by dividing it by 4.7 at each iteration. You will recognize this as a crude round-off of the Feigenbaum number.

Figure 5-4. Bifurcation Diagram of x(1 - x)

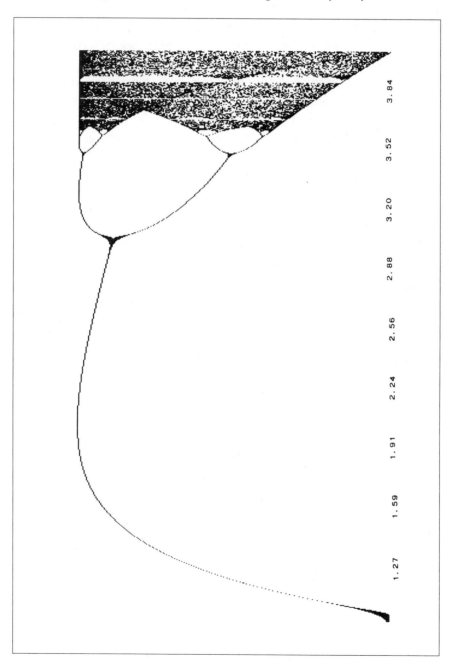

Figure 5-5. Expansion of the Bifurcation Diagram of x(1 - x)

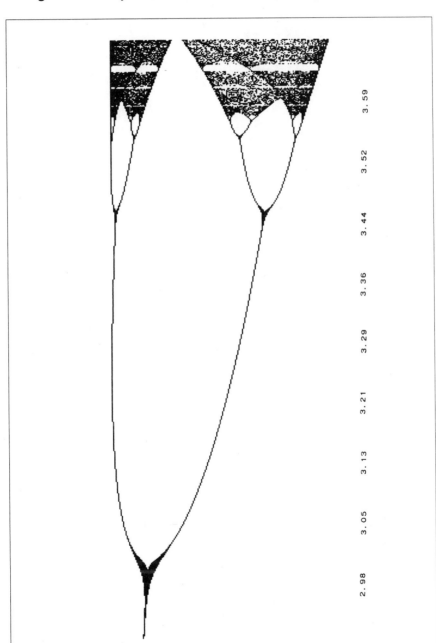

Figure 5-6. Program to Generate Roots of Population Equation

```
program feigenbm;

uses CRT;

var
    x,lambda,f,step_size,old_x,test,lambda_1,lambda_2,delta,
    init_step,old_lambda,new_step,old_step: extended;

    i,iterations: longint;

    j,sign: integer;

begin
    ClrScr;
    lambda := 3.0;
      Writeln('n          Lambda            Delta');
    init_step := 1;
    for j:=1 to 19 do
    begin
        if j mod 2 = 0 then
            sign := -1
        else
            sign := 1;
        GotoXY(0,15+j);
          init_step := init_step/4.67;
          step_size := init_step;
          iterations := 2;
          for i:=1 to j-1 do
           iterations := iterations*2;
          old_x := 0.5;
          lambda := lambda + step_size;
          repeat
          x := old_x;
          for i:=0 to iterations-1 do
             x := lambda*x*(1-x);
           test := (x - old_x)*sign;
            if test < 0 then
              begin
                lambda := lambda - step_size;
                step_size := step_size/2;
              end;
            old_lambda := lambda;
            lambda := lambda + step_size;
              GotoXY(1,j+3);
           write(j:2,'   ',lambda:20:18,'   ');
          until old_lambda >= lambda;
          if j > 2 then
        begin
```

```
          delta := (lambda_1 - lambda_2)/(lambda - lambda_1);
          write(delta:20:18);
        end;
            lambda_2 := lambda_1;
            lambda_1 := lambda;
    end;
end.
```

Figure 5-7. Values of Roots and Feigenbaum Number

n	r	delta
1	3.23606797749978969	
2	3.49856169932770152	
3	3.55464086276882486	4.68077099801069546
4	3.56666737985626851	4.66295961111410222
5	3.56924353163711033	4.66840392591840145
6	3.56979529374994462	4.66895374096762252
7	3.56991346542234851	4.66915718132887754
8	3.56993877423330548	4.66919100248498318
9	3.56994419460806493	4.66919947054711264
10	3.56994535548646858	4.66920113460536986
11	3.56994560411107844	4.66920150943092950
12	3.56994565735885649	4.66920158824756554
13	3.56994566876289996	4.66920160286821290
14	3.56994567120529684	4.66920162588405047
15	3.56994567172838347	4.66920155028815866
16	3.56994567184041260	4.66920176639898966
17	3.56994567186440580	4.66920254147147496
18	3.56994567186954440	4.66921926366696325
19	3.56994567187064489	4.66934028643927598

6

The Snowflake and
Other von Koch Curves

The next few chapters will discuss fractals which are generated using a recursive initiator/generator technique which results in complete self-similarity. Their similarity dimension is the same as their fractal and Hausdorff-Besicovitch dimensions and is easily defined as discussed in Chapter 2. Such curves are constructed using the following technique. We start with an initiator, which may be a straight line or a polygon. Each side of the initiator is then replaced by a generator, which is a connected set of straight lines which form a path from the beginning to the end of the line being replaced. (Usually the points of the generator are on a square grid or a grid made up of equilateral triangles.) Then, each straight line segment of the new figure is replaced by a scaled down version of the generator. This process continues indefinitely. Of course, in reality, we cannot continue the process an infinite number of times, and even if we did, the result would not be interesting, since the detail would be far beyond the resolution of our computer monitor. In practice, we perform from two to sixteen repetitions.

The von Koch Snowflake

This figure was first constructed by the mathematician Helge von Koch in 1904. The initiator, shown in Figure 6-1(a), is an equilateral triangle. The generator, shown in Figure 6-1(b), divides each line segment into three equal parts. Each segment of the generator has a length r of 1/3. The first segment of the generator follows the original line segment. The next two segments form the two sides of an equilateral triangle, the base of which is the second third of the original line. Finally, the fourth segment is identical with the final third of the original line. Thus, the number of segments of the generator, N, is 4. From equation 2-1 of Chapter 2, we find the fractal (or similarity) dimension of the snowflake to be:

```
D = log N / log (1/r) = log 4 / log 3 = 1.2618        (Equation 6-1)
```

Figure 6-2 shows the resulting snowflake for two, three, four, and six iterations.

Figure 6-1. Initiator and Generator for Snowflake

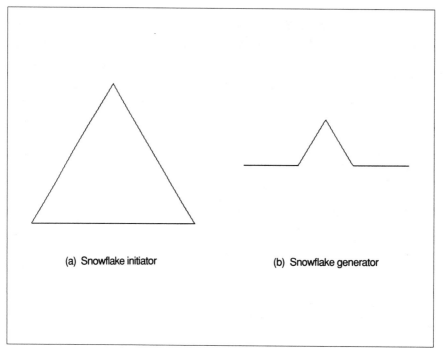

(a) Snowflake initiator   (b) Snowflake generator

Figure 6-2. von Koch Snowflakes with 2, 3, 4, and 6 levels

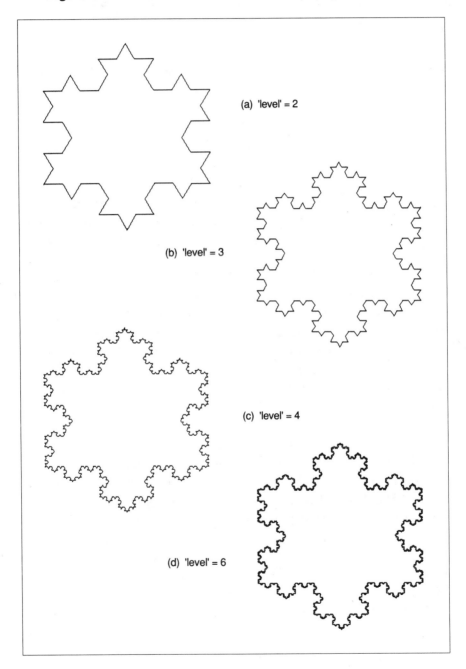

(a) 'level' = 2

(b) 'level' = 3

(c) 'level' = 4

(d) 'level' = 6

Generic Initiator/Generator Program

Figure 6-3 is a program to generate the von Koch snowflake. It can be considered a somewhat generic program for creating this type of curve. As we progress through the next few chapters, we will encounter various complications as the generators become more complex, and we will learn how to deal with them.

For each time that we replace a line segment by the generator, we are going to create an array of coordinate locations (stored in the array *Xpoints* for the x coordinates, and in the array *Ypoints* for the y coordinates and then draw a line from the first set of coordinates to the second, from the second to the third, and so forth until we have drawn as many line segments as are specified by the parameter *generator_size*. To generate these coordinate pairs, we will make use of the modified turtle graphics commands which were developed in Chapter 3. We first identify the beginning and end of the line segment and store the coordinates of each as the beginning and end points of our coordinate arrays. Then we insert these values into the function point which sets the turtle direction (*turtle_theta*) along the original line segment. The step size for turtle movement (*turtle_r*) is determined by measuring the length of the line segment and dividing by the proper divisor to get *r*.

Note that for the snowflake this divisior is 3; we will see later that it can take on other values for other curves. Next, we use *turn*, if necessary, to properly position the turtle. We then use *step* to advance the turtle and record its new position in the position arrays. At any time in the process of stepping through the pattern for the generator, we can record the turtle position in any set of members of the coordinate arrays. Thus, the turtle does not have to follow the actual path which makes up the generator, as long as it stops at every pair of endpoints for the generator lines. We can store every pair of coordinates that are needed in the proper location, regardless of when it was generated, so we have considerable flexibility as to how we are going to create the generator. Our turtle graphics system makes use of a coordinate system that has the point (0, 0) at the center of the display screen. The limits in the x direction are -319 and +320, and the limits in the y direction are -239 (at the bottom of the screen) and +240 at the top.

Figure 6-3. Program to Generate von Koch Snowflake

```
program snoflake;

uses CRT,Graph,Fractal;

var

    graphDriver,GraphMode,i,generator_size,level,init_size:
        integer;

    Xpoints, Ypoints: array[0..24] of real;
    initiator_x1,initiator_x2,initiator_y1,initiator_y2:
        array[0..9] of real;
    ch: char;

procedure generate (X1: real; Y1: real; X2: real; Y2: real;
    level: integer);
    var
        j,k: integer;
        a, b: real;
        Xpoints, Ypoints: array[0..24] of real;

    begin
        dec(level);
        turtle_r := (sqrt((X2 - X1)*(X2 - X1) + (Y2 -
            Y1)*(Y2 - Y1)))/3.0;
        Xpoints[0] := X1;
        Ypoints[0] := Y1;
        Xpoints[4] := X2;
        Ypoints[4] := Y2;
        turtle_theta := point(X1,Y1,X2,Y2);
        turtle_x := X1;
        turtle_y := Y1;
        step;
        Xpoints[1] := turtle_x;
        Ypoints[1] := turtle_y;
        turn(60);
        step;
        Xpoints[2] := turtle_x;
        Ypoints[2] := turtle_y;
        turn(-120);
        step;
```

```
            Xpoints[3] := turtle_x;
            Ypoints[3] := turtle_y;
            if level > 0 then
            begin
                for j:=0 to generator_size - 2 do
                begin
                    X1 := Xpoints[j];
                     X2 := Xpoints[j+1];
                     Y1 := Ypoints[j];
                     Y2 := Ypoints[j+1];
                    generate (X1,Y1,X2,Y2,level);
                end;
            end
            else
            begin
                for k:=0 to generator_size - 2 do
                begin
                    Line(Round(Xpoints[k]+320),Round(175 -
                        Ypoints[k]*0.729),Round(Xpoints[k+1]
                        + 320),Round(175 - Ypoints[k+1]*
                        0.729));
                end;
            end;
        end;
    end;

begin

    generator_size := 5;
    init_size := 3;
    initiator_x1[0] := -150;
    initiator_x1[1] := 0;
    initiator_x1[2] := 150;
    initiator_x2[0] := 0;
    initiator_x2[1] := 150;
    initiator_x2[2] := -150;
    initiator_y1[0] := -75;
    initiator_y1[1] := 185;
    initiator_y1[2] := -75;
    initiator_y2[0] := 185;
    initiator_y2[1] := -75;
    initiator_y2[2] := -75;
    write('Enter level (1 - 8): ');
    readln(level);
    if level < 1 then
        level := 1;
    GraphDriver := 4;
    GraphMode := EGAHi;
    InitGraph(graphDriver,GraphMode,'');
    SetLineStyle(0,$FFFF,1);
    SetColor(15);
    for i:=0 to init_size - 1 do
```

```
  begin
      generate(initiator_x1[i], initiator_y1[i],
          initiator_x2[i], initiator_y2[i], level);
  end;
  ch := ReadKey;
end.
```

The *Line* graphics procedure from Turbo Pascal requires inputs in display coordinates, which are integers. This requires that we apply conversion factors to the turtle coordinates and then use the *Round* procedure to convert the floating point converted turtle coordinates to integers before we actually draw each line. The von Koch snowflake curve generator is so simple that all we have to do is trace its path with the turtle.

The main part of the program allows the user to enter *level* which determines how many recursions will be used to generate the figure, and then calls *generate* for each line segment of the initiator. The *generate* function decrements *level* and then determines the coordinates of all points needed to draw the generator in place of the line segment whose beginning and end points passed as parameters to the function. Then, if the level is greater than 0, the function starts a *for* loop which determines the beginning and end points of each new line segment in the array of points just created by the turtle functions. Then the function calls *generate* to replace each line segment by a new generator. You should note that the *Xpoints* and *Ypoints* arrays are not global, so that each time *generate* is called, a new pair of coordinate arrays is created. Thus there can be quite a few coordinate arrays if *level* is set to a large number.

When *level* is finally decremented to zero, the function actually draws the line segments that are specified by the coordinate arrays at that time and there is no more recursion, so that the program returns to the previous level and continues, until all of the *for* loops have been completed.

The Gosper Curve

This variation of the von Koch curve was discovered by W. Gosper. The initiator is a regular hexagon, and the generator consists of three segments on a grid of equilateral triangles. This and the next curve are a little peculiar in that the line segment to be replaced does not lie on any of the grid lines. Remembering that the turtle point function points the turtle in the direction of the line segment, if you were writing a program to draw this curve you would have to compute the angle that the first piece of the generator makes with the line segment and turn the turtle in this direction before taking the first step. The program listings have already done that for you in these two curves.

Figure 6-4 shows the initiator and the generator laid out on the grid. Applying a little simple geometry shows that if the length from one end of the generator to the other is taken to be one, the length of each of the three segments is:

$$r = 1/\sqrt{7} \qquad\qquad \text{Equation 6-2}$$

Since N = 3, the fractal dimension of the Gosper curve is:

$$D = \log 3 \ / \ \log (\sqrt{7}) = 1.1292 \qquad \text{Equation 6-3}$$

Figure 6-5 shows the resulting curve for level equal to 1, 2, 4, and 6. The program to generate this curve is given in Figure 6-6. It is the same as the snowflake program except for the change in the values for the initiator and the modification of the generate function.

Figure 6-4. Initiator and Generator for Gosper Curve

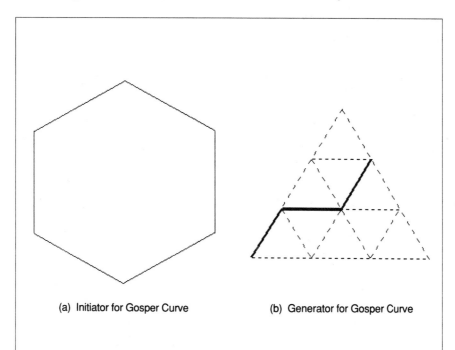

(a) Initiator for Gosper Curve   (b) Generator for Gosper Curve

Figure 6-5. Gosper Curves for levels 1 to 6

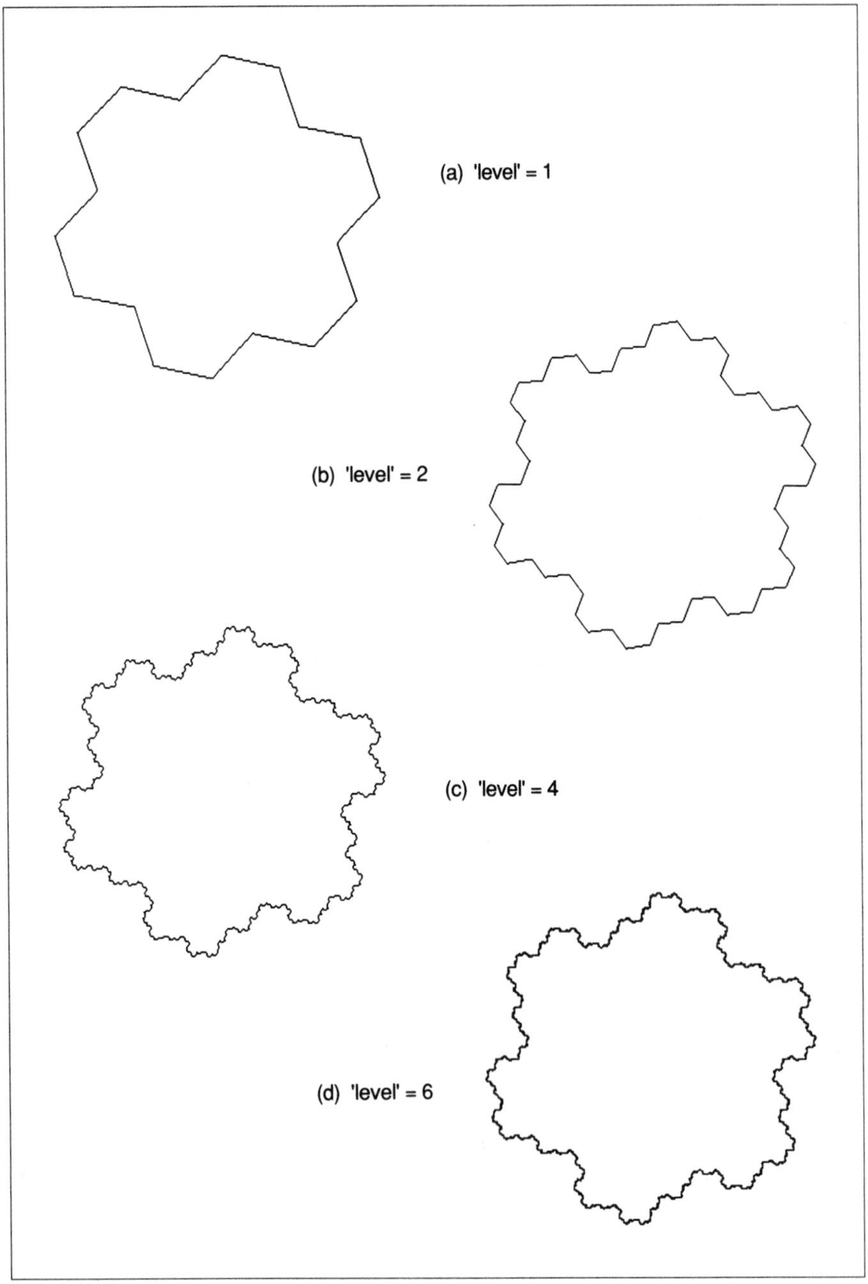

(a) 'level' = 1

(b) 'level' = 2

(c) 'level' = 4

(d) 'level' = 6

Figure 6-6. Program to Generate Gosper Curves

```
program gosper;

uses CRT,Graph,Fractal;

var

    graphDriver,GraphMode,i,generator_size,level,init_size:
        integer;

    Xpoints, Ypoints: array[0..24] of real;
    initiator_x1,initiator_x2,initiator_y1,initiator_y2:
        array[0..9] of real;
    ch: char;


procedure generate (X1: real; Y1: real; X2: real; Y2: real;
    level: integer);
    var
        j,k: integer;
        a, b: real;
        Xpoints, Ypoints: array[0..24] of real;

    begin

        dec(level);
        turtle_r := sqrt(((X2 - X1)*(X2 - X1) + (Y2 -
            Y1)*(Y2 - Y1))/7.0);
        Xpoints[0] := X1;
        Ypoints[0] := Y1;
        Xpoints[3] := X2;
        Ypoints[3] := Y2;
        turtle_theta := point(X1,Y1,X2,Y2);
        turtle_x := X1;
        turtle_y := Y1;
        turn(19.1);
        step;
        Xpoints[1] := turtle_x;
        Ypoints[1] := turtle_y;
        turn(-60);
        step;
        Xpoints[2] := turtle_x;
        Ypoints[2] := turtle_y;
        if level > 0 then
        begin
            for j:=0 to generator_size - 1 do
            begin
                X1 := Xpoints[j];
                X2 := Xpoints[j+1];
```

```
                    Y1 := Ypoints[j];
                    Y2 := Ypoints[j+1];
                  generate (X1,Y1,X2,Y2,level);
              end;
        end
        else
        begin
            for k:=0 to generator_size - 1 do
            begin
                Line(Round(Xpoints[k]+320),Round(175 -
                    Ypoints[k]*0.729)
                    ,Round(Xpoints[k+1]+320),Round(175
                    - Ypoints[k+1]*
                    0.729));
            end;
        end;
    end;

begin

    generator_size := 3;
    init_size := 6;
    initiator_x1[0] := 0;
    initiator_x1[1] := 130;
    initiator_x1[2] := 130;
    initiator_x1[3] := 0;
    initiator_x1[4] := -130;
    initiator_x1[5] := -130;
    initiator_x2[0] := 130;
    initiator_x2[1] := 130;
    initiator_x2[2] := 0;
    initiator_x2[3] := -130;
    initiator_x2[4] := -130;
    initiator_x2[5] := 0;
    initiator_y1[0] := 150;
    initiator_y1[1] := 75;
    initiator_y1[2] := -75;
    initiator_y1[3] := -150;
    initiator_y1[4] := -75;
    initiator_y1[5] := 75;
    initiator_y2[0] := 75;
    initiator_y2[1] := -75;
    initiator_y2[2] := -150;
    initiator_y2[3] := -75;
    initiator_y2[4] := 75;
    initiator_y2[5] := 150;
    write('Enter level (1 - 8): ');
    readln(level);
    if level < 1 then
        level := 1;
    GraphDriver := 4;
```

```
GraphMode := EGAHi;
InitGraph(graphDriver,GraphMode,'');
SetLineStyle(0,$FFFF,1);
SetColor(15);
for i:=0 to init_size - 1 do
begin
    generate(initiator_x1[i], initiator_y1[i],
        initiator_x2[i], initiator_y2[i], level);
end;
ch := ReadKey;
end.
```

Three Segment Quadric von Koch Curve

The next few curves are called "quadric" because the initiator is a square. However, there is nothing sacred about the square initiator; it could be any regular polygon or some other weird figure. An example will be given later. Furthermore, we are going to create our generators on a square grid. For the first of these curves, a three segment generator will be used; N is the same as for the previous curve, but because of the square grid, the length of a segment is:

$$r = 1/\sqrt{5} \qquad\qquad \text{Equation 6-4}$$

and the fractal dimension is different,

$$D = \log 3 / \log(\sqrt{5}) = 1.3652 \qquad \text{Equation 6-5}$$

Figure 6-7 shows the initiator and generator and Figure 6-8 shows the curve for various levels. Again the generic program is used, with appropriate modification to the generate function. The resulting program is listed in Figure 6-9.

**Figure 6-7. Initiator and Generator for the
Three Segment von Koch Curve**

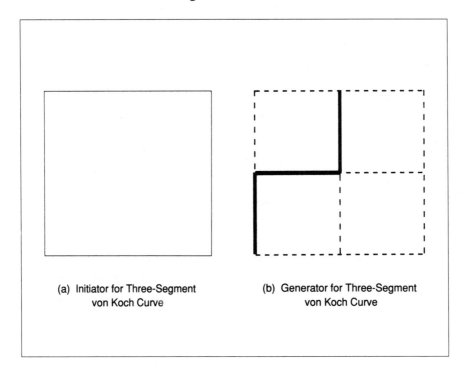

(a) Initiator for Three-Segment
von Koch Curve

(b) Generator for Three-Segment
von Koch Curve

Figure 6-8. Three Segment von Koch Curves—levels 1 to 6

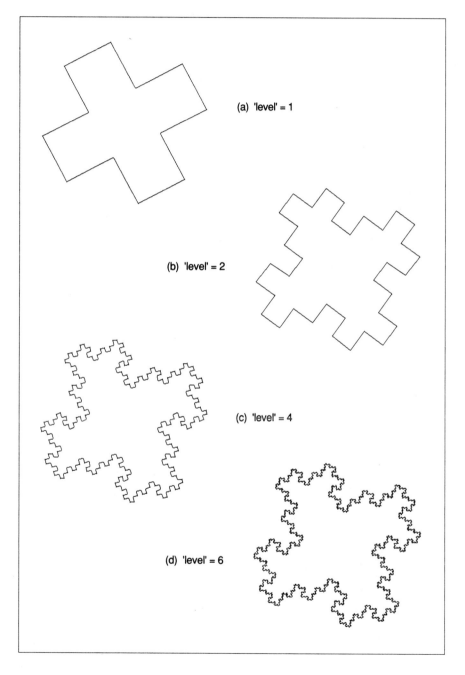

(a) 'level' = 1

(b) 'level' = 2

(c) 'level' = 4

(d) 'level' = 6

Figure 6-9. Program to Generate Three Segment von Koch Curves

```
program qkoch3;

uses CRT,Graph,Fractal;

var

    graphDriver,GraphMode,i,generator_size,level,init_size:
        integer;

    Xpoints, Ypoints: array[0..24] of real;
    initiator_x1,initiator_x2,initiator_y1,initiator_y2:
        array[0..9] of real;
    ch: char;

procedure generate (X1: real; Y1: real; X2: real; Y2: real;
    level: integer);
    var
        j,k: integer;
        a, b: real;
        Xpoints, Ypoints: array[0..24] of real;

    begin

        dec(level);
        turtle_r := sqrt((((X2 - X1)*(X2 - X1) + (Y2 -
            Y1)*(Y2 - Y1))/5.0);
        Xpoints[0] := X1;
        Ypoints[0] := Y1;
        Xpoints[3] := X2;
        Ypoints[3] := Y2;
        turtle_theta := point(X1,Y1,X2,Y2);
        turtle_x := X1;
        turtle_y := Y1;
        turn(26.56);
        step;
        Xpoints[1] := turtle_x;
        Ypoints[1] := turtle_y;
        turn(-90);
        step;
        Xpoints[2] := turtle_x;
        Ypoints[2] := turtle_y;
        if level > 0 then
        begin
            for j:=0 to generator_size - 1 do
            begin
                X1 := Xpoints[j];
                X2 := Xpoints[j+1];
```

```
                Y1 := Ypoints[j];
                Y2 := Ypoints[j+1];
              generate (X1,Y1,X2,Y2,level);
          end;
      end
      else
      begin
          for k:=0 to generator_size - 1 do
          begin
              Line(Round(Xpoints[k]+320),Round(175 -
                  Ypoints[k]*0.729)
                  ,Round(Xpoints[k+1]+320),Round(175
                  - Ypoints[k+1]*
                  0.729));
          end;
      end;
  end;

begin

    generator_size := 3;
    init_size := 4;
    initiator_x1[0] := -130;
    initiator_x1[1] := 130;
    initiator_x1[2] := 130;
    initiator_x1[3] := -130;
    initiator_x2[0] := 130;
    initiator_x2[1] := 130;
    initiator_x2[2] := -130;
    initiator_x2[3] := -130;
    initiator_y1[0] := 130;
    initiator_y1[1] := 130;
    initiator_y1[2] := -130;
    initiator_y1[3] := -130;
    initiator_y2[0] := 130;
    initiator_y2[1] := -130;
    initiator_y2[2] := -130;
    initiator_y2[3] := 130;
    write('Enter level (1 - 8): ');
    readln(level);
    if level < 1 then
        level := 1;
    GraphDriver := 4;
    GraphMode := EGAHi;
    InitGraph(graphDriver,GraphMode,'');
    SetLineStyle(0,$FFFF,1);
    SetColor(15);
    for i:=0 to init_size - 1 do
    begin
        generate(initiator_x1[i], initiator_y1[i],
            initiator_x2[i], initiator_y2[i], level);
```

```
      end;
      ch := ReadKey;
end.
```

Eight Segment Quadric von Koch Curve

The next few curves are all going to make use of a square grid and turning angles of ninety degrees. They are a little more regular than the previous curve because the line segment to be replaced falls along the middle horizontal line of the grid. For the first curve to be considered, we will let:

$$r = 1/4 \qquad\qquad \text{(Equation 6-6)}$$

We can now draw various generators, the only limitation being that we want the curve to have no self-overlap and no self-intersection. If we also want the curve to have the highest fractal dimension possible, we need to find the generator for which N is the largest. Mandelbrot states that the highest possible value of N is:

$$N_{max} = 1/2r^2 \qquad\qquad \text{(Equation 6-7)}$$

when r is even and

$$N_{max} = (1 + r^2)/2r^2 \qquad\qquad \text{(Equation 6-8)}$$

when r is odd. Thus, for $r = 1/4$, we find that N_{max} is 8. The fractal dimension of this curve is thus:

$$D = \log 8 \,/\, \log 4 = 1.5 \qquad \text{(Equation 6-9)}$$

Figure 6-10 shows the initiator and generator for this curve and Figure 6-11 shows the curve for levels of 1, 2, 4, and 6. The program to generate this curve is listed in Figure 6-12.

Figure 6-10. Initiator and Generator for Eight Segment von Koch Curve

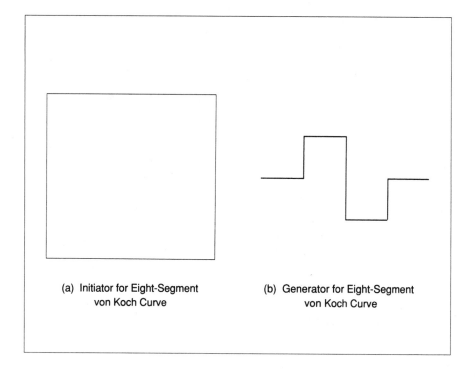

(a) Initiator for Eight-Segment
von Koch Curve

(b) Generator for Eight-Segment
von Koch Curve

Figure 6-11. Eight Segment von Koch Curves-levels 1 to 6

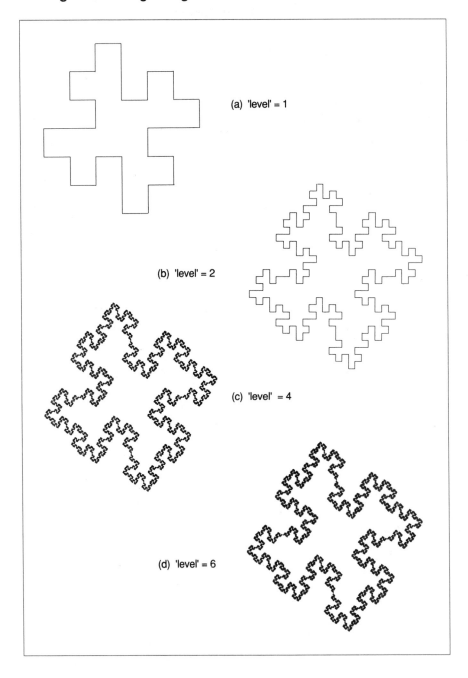

(a) 'level' = 1

(b) 'level' = 2

(c) 'level' = 4

(d) 'level' = 6

Figure 6-12. Program to Generate Eight Segment von Koch Curves

```
program qkoch8;

uses CRT,Graph,Fractal;

var

    graphDriver,GraphMode,i,generator_size,level,init_size:
integer;

    Xpoints, Ypoints: array[0..24] of real;
    initiator_x1,initiator_x2,initiator_y1,initiator_y2:
        array[0..9] of real;
    ch: char;


procedure generate (X1: real; Y1: real; X2: real; Y2: real;
    level: integer);
    var
        j,k: integer;
        a, b: real;
        Xpoints, Ypoints: array[0..24] of real;

    begin

        dec(level);
        turtle_r := sqrt((X2 - X1)*(X2 - X1) + (Y2 - Y1)*(Y2
            - Y1))/4.0;
        Xpoints[0] := X1;
        Ypoints[0] := Y1;
        Xpoints[8] := X2;
        Ypoints[8] := Y2;
        turtle_theta := point(X1,Y1,X2,Y2);
        turtle_x := X1;
        turtle_y := Y1;
        step;
        Xpoints[1] := turtle_x;
        Ypoints[1] := turtle_y;
        turn(90);          step;
        Xpoints[2] := turtle_x;
        Ypoints[2] := turtle_y;
        turn(-90);
        step;
        Xpoints[3] := turtle_x;
        Ypoints[3] := turtle_y;
        turn(-90);
        step;
        Xpoints[4] := turtle_x;
```

```
            Ypoints[4] := turtle_y;
            step;
            Xpoints[5] := turtle_x;
            Ypoints[5] := turtle_y;
            turn(90);
            step;
            Xpoints[6] := turtle_x;
            Ypoints[6] := turtle_y;
            turn(90);
            step;
            Xpoints[7] := turtle_x;
            Ypoints[7] := turtle_y;
            if level > 0 then
            begin
                for j:=0 to generator_size - 1 do
                begin
                    X1 := Xpoints[j];
                      X2 := Xpoints[j+1];
                      Y1 := Ypoints[j];
                      Y2 := Ypoints[j+1];
                    generate (X1,Y1,X2,Y2,level);
                end;
            end
            else
            begin
                for k:=0 to generator_size - 1 do
                begin
                    Line(Round(Xpoints[k]+320),Round(175 -
                        Ypoints[k]*0.729)
                        ,Round(Xpoints[k+1]+320),Round(175
                        - points[k+1]* 0.729));
                end;
            end;
      end;

begin

    generator_size := 8;
    init_size := 4;
    initiator_x1[0] := -130;
    initiator_x1[1] := 130;
    initiator_x1[2] := 130;
    initiator_x1[3] := -130;
    initiator_x2[0] := 130;
    initiator_x2[1] := 130;
    initiator_x2[2] := -130;
    initiator_x2[3] := -130;
    initiator_y1[0] := 130;
    initiator_y1[1] := 130;
    initiator_y1[2] := -130;
    initiator_y1[3] := -130;
```

```
    initiator_y2[0] := 130;
    initiator_y2[1] := -130;
    initiator_y2[2] := -130;
    initiator_y2[3] := 130;
    write('Enter level (1 - 8): ');
    readln(level);
    if level < 1 then
        level := 1;
    GraphDriver := 4;
    GraphMode := EGAHi;
    InitGraph(graphDriver,GraphMode,'');
    SetLineStyle(0,$FFFF,1);
    SetColor(15);
    for i:=0 to init_size - 1 do
    begin
        generate(initiator_x1[i], initiator_y1[i],
            initiator_x2[i], initiator_y2[i], level);
    end;
    ch := ReadKey;
end.
```

Eighteen Segment Quadric von Koch Curve

If we let

$$r = 1/6 \qquad \text{(Equation 6-10)}$$

we find that N_{max} is 18. The fractal dimension of this curve is:

$$D = \log 18 \ / \ \log 6 = 1.6131 \qquad \text{(Equation 6-11)}$$

Figure 6-13 shows the initiator and generator for this curve, and Figure 6-14 shows the curve for levels of 1, 2, 3, and 4. The program to generate this curve is listed in Figure 6-15.

**Figure 6-13. Initiator and Generator for 18 Segment
von Koch Curve**

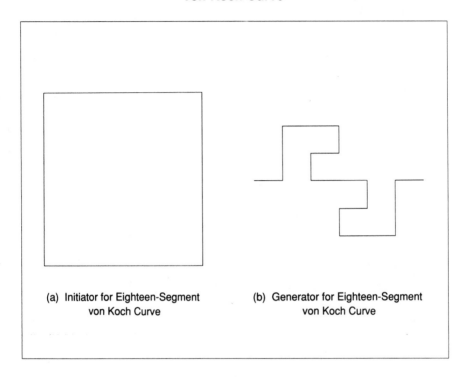

(a) Initiator for Eighteen-Segment
von Koch Curve

(b) Generator for Eighteen-Segment
von Koch Curve

Figure 6-14. 18 Segment von Koch Curves-levels 1 to 4

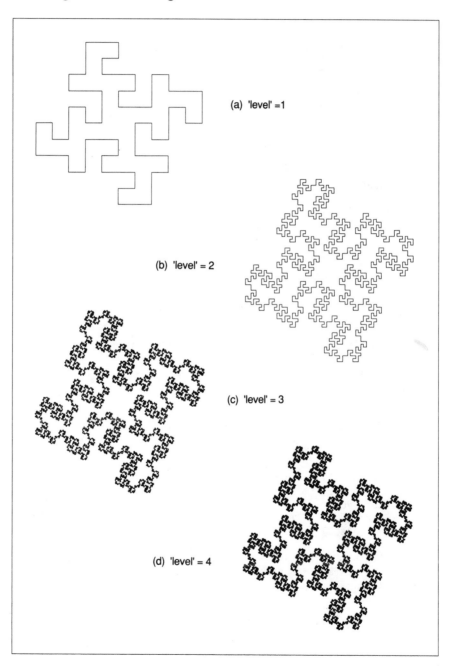

(a) 'level' =1

(b) 'level' = 2

(c) 'level' = 3

(d) 'level' = 4

Figure 6-15. Program to Generate 18 Segment von Koch Curves

```
program qkoch18;

uses CRT,Graph,Fractal;

var

    graphDriver,GraphMode,i,generator_size,level,init_size:
        integer;

    Xpoints, Ypoints: array[0..24] of real;
    initiator_x1,initiator_x2,initiator_y1,initiator_y2:
        array[0..9] of real;
    ch: char;

procedure generate (X1: real; Y1: real; X2: real; Y2: real;
    level: integer);
    var
        j,k: integer;
        a, b: real;
        Xpoints, Ypoints: array[0..24] of real;

    begin

        dec(level);
        turtle_r := sqrt((X2 - X1)*(X2 - X1) + (Y2 - Y1)*(Y2
            - Y1))/6.0;
        Xpoints[0] := X1;
        Ypoints[0] := Y1;
        Xpoints[18] := X2;
        Ypoints[18] := Y2;
        turtle_theta := point(X1,Y1,X2,Y2);
        turtle_x := X1;
        turtle_y := Y1;
        step;
        Xpoints[1] := turtle_x;
        Ypoints[1] := turtle_y;
        turn(90);
        step;           Xpoints[2] := turtle_x;
        Ypoints[2] := turtle_y;
        step;
        Xpoints[3] := turtle_x;
        Ypoints[3] := turtle_y;
        turn(-90);
        step;
        Xpoints[4] := turtle_x;
        Ypoints[4] := turtle_y;
        step;
        Xpoints[5] := turtle_x;
```

```
Ypoints[5] := turtle_y;
turn(-90);
step;
Xpoints[6] := turtle_x;
Ypoints[6] := turtle_y;
turn(-90);
step;
Xpoints[7] := turtle_x;
Ypoints[7] := turtle_y;
turn(90);
step;
Xpoints[8] := turtle_x;
Ypoints[8] := turtle_y;
turn(90);
step;
Xpoints[9] := turtle_x;
Ypoints[9] := turtle_y;
step;
Xpoints[10] := turtle_x;
Ypoints[10] := turtle_y;
turn(-90);
step;
Xpoints[11] := turtle_x;
Ypoints[11] := turtle_y;
turn(-90);
step;
Xpoints[12] := turtle_x;
Ypoints[12] := turtle_y;
turn(90);
step;
Xpoints[13] := turtle_x;
Ypoints[13] := turtle_y;
turn(90);
step;
Xpoints[14] := turtle_x;
Ypoints[14] := turtle_y;
step;
Xpoints[15] := turtle_x;
Ypoints[15] := turtle_y;
turn(90);
step;
Xpoints[16] := turtle_x;
Ypoints[16] := turtle_y;
step;
Xpoints[17] := turtle_x;
Ypoints[17] := turtle_y;
if level > 0 then
begin
    for j:=0 to generator_size - 1 do
    begin
        X1 := Xpoints[j];
```

```
                        X2 := Xpoints[j+1];
                        Y1 := Ypoints[j];
                        Y2 := Ypoints[j+1];
                    generate (X1,Y1,X2,Y2,level);
                end;
        end
        else
        begin
            for k:=0 to generator_size - 1 do
            begin
                Line(Round(Xpoints[k]+320),Round(175 -
                Ypoints[k]*0.729)
                ,Round(Xpoints[k+1]+320),Round(175 -
                Ypoints[k+1]* 0.729));
            end;
        end;
    end;

begin
    generator_size := 18;
    init_size := 4;
    initiator_x1[0] := -130;
    initiator_x1[1] := 130;
    initiator_x1[2] := 130;
    initiator_x1[3] := -130;
    initiator_x2[0] := 130;
    initiator_x2[1] := 130;
    initiator_x2[2] := -130;
    initiator_x2[3] := -130;
    initiator_y1[0] := 130;
    initiator_y1[1] := 130;
    initiator_y1[2] := -130;
    initiator_y1[3] := -130;
    initiator_y2[0] := 130;
    initiator_y2[1] := -130;
    initiator_y2[2] := -130;
    initiator_y2[3] := 130;
    write('Enter level (1 - 8): ');
    readln(level);
    if level < 1 then
        level := 1;
    GraphDriver := 4;
    GraphMode := EGAHi;
    InitGraph(graphDriver,GraphMode,'');
    SetLineStyle(0,$FFFF,1);
    SetColor(15);
    for i:=0 to init_size - 1 do
    begin
        generate(initiator_x1[i], initiator_y1[i],
            initiator_x2[i], initiator_y2[i], level);
    end;
```

```
    ch := ReadKey;
end.
```

Thirty-two Segment Quadric von Koch Curve

If we let

$$r = 1/8 \qquad \text{(Equation 6-12)}$$

we find that N_{max} is 32. The fractal dimension of this curve is:

$$D = \log 32 \ / \ \log 8 = 1.6667 \qquad \text{(Equation 6-13)}$$

Figure 6-16 shows the initiator and generator for this curve and Figure 6-17 shows the curve for levels of 1, 2, and 3. The program to generate this curve is is listed in Figure 6-18.

**Figure 6-16. Initiator and Generator for 32 Segment
von Koch Curve**

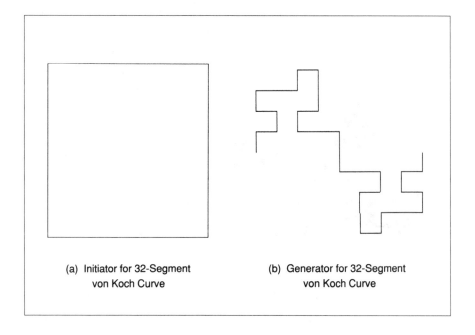

(a) Initiator for 32-Segment
von Koch Curve

(b) Generator for 32-Segment
von Koch Curve

Figure 6-17. 32 Segment von Koch Curves-levels 1 to 3

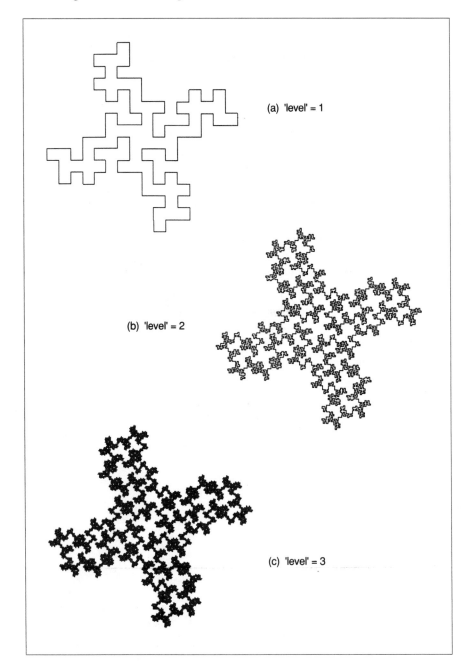

(a) 'level' = 1

(b) 'level' = 2

(c) 'level' = 3

Figure 6-18 Program to Generate 32 Segment von Koch Curves

```
program qkoch32;

uses CRT,Graph,Fractal;

var

    graphDriver,GraphMode,i,generator_size,level,init_size:
        integer;

    Xpoints, Ypoints: array[0..24] of real;
    initiator_x1,initiator_x2,initiator_y1,initiator_y2:
        array[0..9] of real;
    ch: char;

procedure generate (X1: real; Y1: real; X2: real; Y2: real;
    level: integer);
    var
        j,k: integer;
        a, b: real;
        Xpoints, Ypoints: array[0..32] of real;

    begin

        dec(level);
        turtle_r := sqrt((X2 - X1)*(X2 - X1) + (Y2 - Y1)*(Y2
            - Y1))/8.0;
        Xpoints[0] := X1;
        Ypoints[0] := Y1;
        Xpoints[32] := X2;
        Ypoints[32] := Y2;
        turtle_theta := point(X1,Y1,X2,Y2);
        turtle_x := X1;
        turtle_y := Y1;
        turn(90);
        step;
        Xpoints[1] := turtle_x;
        Ypoints[1] := turtle_y;
        turn(-90);
        step;
        Xpoints[2] := turtle_x;
        Ypoints[2] := turtle_y;
        turn(90);
        step;
        Xpoints[3] := turtle_x;
        Ypoints[3] := turtle_y;
        turn(90);
        step;
        Xpoints[4] := turtle_x;
```

```
Ypoints[4] := turtle_y;
turn(-90);
step;
Xpoints[5] := turtle_x;
Ypoints[5] := turtle_y;
turn(-90);
step;
Xpoints[6] := turtle_x;
Ypoints[6] := turtle_y;
step;
Xpoints[7] := turtle_x;
Ypoints[7] := turtle_y;
turn(90);
step;
Xpoints[8] := turtle_x;
Ypoints[8] := turtle_y;
turn(-90);
step;
Xpoints[9] := turtle_x;
Ypoints[9] := turtle_y;
turn(-90);
step;
Xpoints[10] := turtle_x;
Ypoints[10] := turtle_y;
step;
Xpoints[11] := turtle_x;
Ypoints[11] := turtle_y;
turn(-90);
step;
Xpoints[12] := turtle_x;
Ypoints[12] := turtle_y;
turn(90);         step;
Xpoints[13] := turtle_x;
Ypoints[13] := turtle_y;
turn(90);
step;
Xpoints[14] := turtle_x;
Ypoints[14] := turtle_y;
step;
Xpoints[15] := turtle_x;
Ypoints[15] := turtle_y;
turn(-90);
step;
Xpoints[16] := turtle_x;
Ypoints[16] := turtle_y;
step;
Xpoints[17] := turtle_x;
Ypoints[17] := turtle_y;
turn(90);
step;
Xpoints[18] := turtle_x;
```

```
Ypoints[18] := turtle_y;
step;
Xpoints[19] := turtle_x;
Ypoints[19] := turtle_y;
turn(-90);
step;
Xpoints[20] := turtle_x;
Ypoints[20] := turtle_y;
turn(-90);
step;
Xpoints[21] := turtle_x;
Ypoints[21] := turtle_y;
turn(90);
step;
Xpoints[22] := turtle_x;
Ypoints[22] := turtle_y;
step;
Xpoints[23] := turtle_x;
Ypoints[23] := turtle_y;
turn(90);
step;
Xpoints[24] := turtle_x;
Ypoints[24] := turtle_y;
turn(90);
step;
Xpoints[25] := turtle_x;
Ypoints[25] := turtle_y;
turn(-90);
step;
Xpoints[26] := turtle_x;
Ypoints[26] := turtle_y;
step;
Xpoints[27] := turtle_x;
Ypoints[27] := turtle_y;
turn(90);
step;
Xpoints[28] := turtle_x;
Ypoints[28] := turtle_y;
turn(90);
step;
Xpoints[29] := turtle_x;
Ypoints[29] := turtle_y;
turn(-90);
step;
Xpoints[30] := turtle_x;
Ypoints[30] := turtle_y;
turn(-90);
step;
Xpoints[31] := turtle_x;
Ypoints[31] := turtle_y;
if level > 0 then
```

```
      begin
          for j:=0 to generator_size - 1 do
          begin
              X1 := Xpoints[j];
              X2 := Xpoints[j+1];
              Y1 := Ypoints[j];
              Y2 := Ypoints[j+1];
              generate (X1,Y1,X2,Y2,level);
          end;
      end
      else           begin
          for k:=0 to generator_size - 1 do
          begin
              Line(Round(Xpoints[k]+320),Round(175 -
                  Ypoints[k]*0.729)
                  ,Round(Xpoints[k+1]+320),Round(175
                  - Ypoints[k+1]* 0.729));
          end;
      end;
   end;

begin

   generator_size := 32;
   init_size := 4;
   initiator_x1[0] := -100;
   initiator_x1[1] := 100;
   initiator_x1[2] := 100;
   initiator_x1[3] := -100;
   initiator_x2[0] := 100;
   initiator_x2[1] := 100;
   initiator_x2[2] := -100;
   initiator_x2[3] := -100;
   initiator_y1[0] := 100;
   initiator_y1[1] := 100;
   initiator_y1[2] := -100;
   initiator_y1[3] := -100;
   initiator_y2[0] := 100;
   initiator_y2[1] := -100;
   initiator_y2[2] := -100;
   initiator_y2[3] := 100;
   write('Enter level (1 - 8): ');
   readln(level);
   if level < 1 then
       level := 1;
   GraphDriver := 4;
   GraphMode := EGAHi;
   InitGraph(graphDriver,GraphMode,'');
   SetLineStyle(0,$FFFF,1);
   SetColor(15);
   for i:=0 to init_size - 1 do       begin
```

```
        generate(initiator_x1[i], initiator_y1[i],
            initiator_x2[i], initiator_y2[i], level);
    end;
    ch := ReadKey;
end.
```

Fifty Segment Quadric von Koch Curve

If we let

$$r = 1/10 \qquad \qquad \text{(Equation 6-14)}$$

we find that N_{max} is 50. The fractal dimension of this curve is:

$$D = \log 50 \; / \; \log 10 = 1.6990 \qquad \text{(Equation 6-15)}$$

As the generator contains more and more segments, it becomes less and less obvious how it is obtained. The process is a sort of trial-and-error one, but at this point, it is time to develop some guide lines for generator creation. Figure 6-19 shows the initiator and generator for the fifty segment curve. The generator grid is also shown. Note that slanting dotted lines have been drawn connecting mid-points of adjacent sides of the grid. If we are to use the generator to replace line segments that meet at ninety-degree angles, we cannot have any part of the generator outside of the bounds of the diamond created by these dotted lines. This is sufficient to avoid self-overlapping, but does not prevent self-intersection. To assure against self-intersection, we mentally merge each pair of parallel sides of the diamond. If the generator touches the diamond side at the same point for both sides of a pair, self-intersection will occur. Finally, the easiest way to create the generator is to create it in two parts that are symmetrical (although possibly a mirror image), each beginning at one end of the line segment being replaced, and ending at its middle. The constraints are thus:

1. Create a half-generator from one end of the line segment to be replaced to its middle, containing $N_{max}/2$ segments.

2.   Do not go outside of the diamond.

3.   If the generator intersects a point on one of a pair of parallel diamond sides, it may not intersect a corresponding point of the other of the pair of sides.

This is where the trial and error comes in. You next seek a path that will contain the required number of segments and meet the above constraints. Once you have the half-generator created, you can turn the graph upside down and draw the same half-generator to complete the process. Figure 6-20 shows the fifty segment curve for levels of 1, 2, and 3. The program to generate this curve is listed in Figure 6-21.

Figure 6-19. Initiator and Generator for 50 Segment von Koch Curve

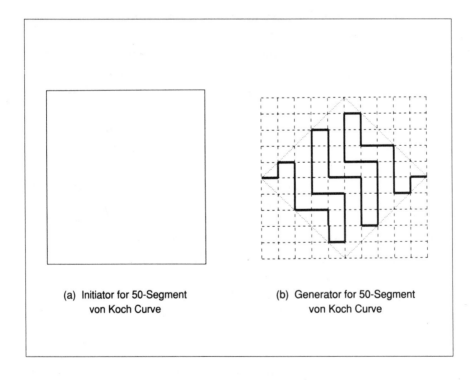

(a) Initiator for 50-Segment
von Koch Curve

(b) Generator for 50-Segment
von Koch Curve

Figure 6-20. 50 Segment von Koch Curves-levels 1 to 3

(a) 'level' = 1

(b) 'level' = 2

(c) 'level' = 3

Figure 6-21. Program to Generate 50 Segment von Koch Curves

```
program qkoch50;

uses CRT,Graph,Fractal;

var

    graphDriver,GraphMode,i,generator_size,level,init_size:
        integer;

    Xpoints, Ypoints: array[0..24] of real;
    initiator_x1,initiator_x2,initiator_y1,initiator_y2:
        array[0..9] of real;
    ch: char;

procedure generate (X1: real; Y1: real; X2: real; Y2: real;
    level: integer);
    var
        j,k: integer;
        a, b: real;
        Xpoints, Ypoints: array[0..50] of real;

    begin

        dec(level);
        turtle_r := sqrt((X2 - X1)*(X2 - X1) + (Y2 - Y1)*(Y2
            - Y1))/10.0;
        Xpoints[0] := X1;
        Ypoints[0] := Y1;
        Xpoints[50] := X2;
        Ypoints[50] := Y2;
        turtle_theta := point(X1,Y1,X2,Y2);
        turtle_x := X1;
        turtle_y := Y1;
        step;
        Xpoints[1] := turtle_x;
        Ypoints[1] := turtle_y;
        turn(90);
        step;
        Xpoints[2] := turtle_x;
        Ypoints[2] := turtle_y;
        turn(-90);
        step;
        Xpoints[3] := turtle_x;
        Ypoints[3] := turtle_y;
        turn(-90);
        step;
        Xpoints[4] := turtle_x;
        Ypoints[4] := turtle_y;
```

```
step;
Xpoints[5] := turtle_x;
Ypoints[5] := turtle_y;
step;
Xpoints[6] := turtle_x;
Ypoints[6] := turtle_y;
turn(90);
step;
Xpoints[7] := turtle_x;
Ypoints[7] := turtle_y;
step;
Xpoints[8] := turtle_x;
Ypoints[8] := turtle_y;
turn(-90);
step;
Xpoints[9] := turtle_x;
Ypoints[9] := turtle_y;
step;
Xpoints[10] := turtle_x;
Ypoints[10] := turtle_y;
turn(90);
step;
Xpoints[11] := turtle_x;
Ypoints[11] := turtle_y;
turn(90);
step;
Xpoints[12] := turtle_x;
Ypoints[12] := turtle_y;
step;
Xpoints[13] := turtle_x;
Ypoints[13] := turtle_y;
step;
Xpoints[14] := turtle_x;
Ypoints[14] := turtle_y;
turn(90);
step;
Xpoints[15] := turtle_x;
Ypoints[15] := turtle_y;
step;
Xpoints[16] := turtle_x;
Ypoints[16] := turtle_y;
turn(-90);
step;
Xpoints[17] := turtle_x;
Ypoints[17] := turtle_y;
step;
Xpoints[18] := turtle_x;
Ypoints[18] := turtle_y;
step;
Xpoints[19] := turtle_x;
Ypoints[19] := turtle_y;
```

```
step;
Xpoints[20] := turtle_x;
Ypoints[20] := turtle_y;
turn(-90);
step;
Xpoints[21] := turtle_x;
Ypoints[21] := turtle_y;
turn(-90);
step;
Xpoints[22] := turtle_x;
Ypoints[22] := turtle_y;
step;
Xpoints[23] := turtle_x;
Ypoints[23] := turtle_y;
step;
Xpoints[24] := turtle_x;
Ypoints[24] := turtle_y;
turn(90);
step;
Xpoints[25] := turtle_x;
Ypoints[25] := turtle_y;
step;
Xpoints[26] := turtle_x;
Ypoints[26] := turtle_y;
turn(-90);
step;
Xpoints[27] := turtle_x;
Ypoints[27] := turtle_y;
step;
Xpoints[28] := turtle_x;
Ypoints[28] := turtle_y;
step;
Xpoints[29] := turtle_x;
Ypoints[29] := turtle_y;
turn(90);
step;
Xpoints[30] := turtle_x;
Ypoints[30] := turtle_y;
turn(90);
step;
Xpoints[31] := turtle_x;
Ypoints[31] := turtle_y;
step;
Xpoints[32] := turtle_x;
Ypoints[32] := turtle_y;
step;
Xpoints[33] := turtle_x;
Ypoints[33] := turtle_y;
step;
Xpoints[34] := turtle_x;
Ypoints[34] := turtle_y;
```

```
turn(90);
step;
Xpoints[35] := turtle_x;
Ypoints[35] := turtle_y;
step;
Xpoints[36] := turtle_x;
Ypoints[36] := turtle_y;
turn(-90);
step;
Xpoints[37] := turtle_x;
Ypoints[37] := turtle_y;
step;
Xpoints[38] := turtle_x;
Ypoints[38] := turtle_y;
step;
Xpoints[39] := turtle_x;
Ypoints[39] := turtle_y;
turn(-90);
step;
Xpoints[40] := turtle_x;
Ypoints[40] := turtle_y;
turn(-90);
step;
Xpoints[41] := turtle_x;
Ypoints[41] := turtle_y;
step;
Xpoints[42] := turtle_x;
Ypoints[42] := turtle_y;
turn(90);
step;
Xpoints[43] := turtle_x;
Ypoints[43] := turtle_y;
step;
Xpoints[44] := turtle_x;
Ypoints[44] := turtle_y;
turn(-90);
step;
Xpoints[45] := turtle_x;
Ypoints[45] := turtle_y;
step;
Xpoints[46] := turtle_x;
Ypoints[46] := turtle_y;
step;
Xpoints[47] := turtle_x;
Ypoints[47] := turtle_y;
turn(90);
step;
Xpoints[48] := turtle_x;
Ypoints[48] := turtle_y;
turn(90);
step;
```

```
        Xpoints[49] := turtle_x;
        Ypoints[49] := turtle_y;
        if level > 0 then
        begin
            for j:=0 to generator_size - 1 do
            begin
                X1 := Xpoints[j];
                X2 := Xpoints[j+1];
                Y1 := Ypoints[j];
                Y2 := Ypoints[j+1];
                generate (X1,Y1,X2,Y2,level);
            end;
        end
        else
        begin
            for k:=0 to generator_size - 1 do
            begin
                Line(Round(Xpoints[k]+320),Round(175 -
                    Ypoints[k]*0.729)
                    ,Round(Xpoints[k+1]+320),Round(175
                    - Ypoints[k+1]* 0.729));
            end;
        end;
    end;

begin

    generator_size := 50;
    init_size := 4;
    initiator_x1[0] := -100;
    initiator_x1[1] := 100;
    initiator_x1[2] := 100;
    initiator_x1[3] := -100;
    initiator_x2[0] := 100;
    initiator_x2[1] := 100;
    initiator_x2[2] := -100;
    initiator_x2[3] := -100;
    initiator_y1[0] := 100;
    initiator_y1[1] := 100;
    initiator_y1[2] := -100;
    initiator_y1[3] := -100;
    initiator_y2[0] := 100;
    initiator_y2[1] := -100;
    initiator_y2[2] := -100;
    initiator_y2[3] := 100;
    write('Enter level (1 - 8): ');
    readln(level);
    if level < 1 then
        level := 1;
    GraphDriver := 4;
    GraphMode := EGAHi;
```

```
        InitGraph(graphDriver,GraphMode,'');
        SetLineStyle(0,$FFFF,1);
        SetColor(15);
        for i:=0 to init_size - 1 do
        begin
            generate(initiator_x1[i], initiator_y1[i],
                initiator_x2[i], initiator_y2[i], level);
        end;
        ch := ReadKey;
end.
```

Using Other Initiators

All of the von Koch curves that have been described above using the square initiator can easily be adapted to other regular polygon initiators of five or more sides. (The generators have been set up so as not to be self-overlapping or self-intersecting as long as the sides of the polygon do not intersect at angles of less than ninety degrees. You can experiment with figures other than regular polygons as long as this condition is met.

Figure 6-22 shows the initiator and generator for an eight segment von Koch curve using a hexagon as the initiator. Figure 6-23 shows the curve for levels of 1, 2, 3, and 4. To generate this curve, all you need to do is run the program for the eight segment von Koch curve, as given above, with the initiator changed to a hexagon. The listing of the resulting program is given in Figure 6-24.

Figure 6-22. Initiator and Generator for Eight Segment von Koch Curve with Hexagonal Generator

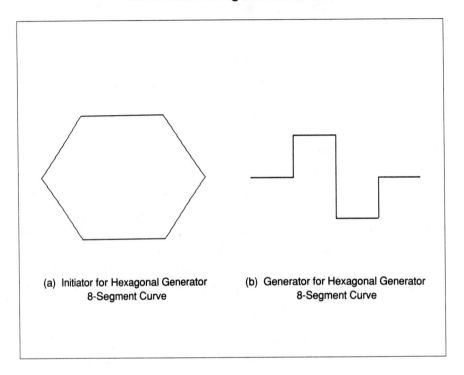

(a) Initiator for Hexagonal Generator
8-Segment Curve

(b) Generator for Hexagonal Generator
8-Segment Curve

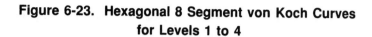

**Figure 6-23. Hexagonal 8 Segment von Koch Curves
for Levels 1 to 4**

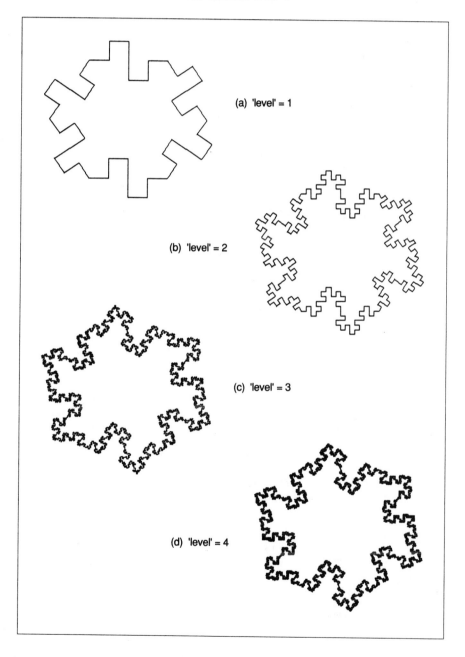

(a) 'level' = 1

(b) 'level' = 2

(c) 'level' = 3

(d) 'level' = 4

Figure 6-24. Program to Generate Hexagonal 8 Segment
von Koch Curve

```
program hkoch8;

uses CRT,Graph,Fractal;

var

    graphDriver,GraphMode,i,generator_size,level,init_size:
        integer;

    Xpoints, Ypoints: array[0..24] of real;
    initiator_x1,initiator_x2,initiator_y1,initiator_y2:
    array[0..9] of real;
    ch: char;

procedure generate (X1: real; Y1: real; X2: real; Y2: real;
    level: integer);
    var
        j,k: integer;
        a, b: real;
        Xpoints, Ypoints: array[0..24] of real;

    begin

        dec(level);
        turtle_r := sqrt((X2 - X1)*(X2 - X1) + (Y2 - Y1)*(Y2
            - Y1))/4.0;
        Xpoints[0] := X1;
        Ypoints[0] := Y1;
        Xpoints[8] := X2;
        Ypoints[8] := Y2;
        turtle_theta := point(X1,Y1,X2,Y2);
        turtle_x := X1;
        turtle_y := Y1;
        step;
        Xpoints[1] := turtle_x;
        Ypoints[1] := turtle_y;
        turn(90);
        step;
        Xpoints[2] := turtle_x;
        Ypoints[2] := turtle_y;
        turn(-90);
        step;
        Xpoints[3] := turtle_x;
        Ypoints[3] := turtle_y;
        turn(-90);
        step;
        Xpoints[4] := turtle_x;
```

```
            Ypoints[4] := turtle_y;
            step;
            Xpoints[5] := turtle_x;
            Ypoints[5] := turtle_y;
            turn(90);
            step;
            Xpoints[6] := turtle_x;
            Ypoints[6] := turtle_y;
            turn(90);
            step;
            Xpoints[7] := turtle_x;
            Ypoints[7] := turtle_y;
            if level > 0 then
            begin
                for j:=0 to generator_size - 1 do
                begin
                    X1 := Xpoints[j];
                      X2 := Xpoints[j+1];
                      Y1 := Ypoints[j];
                      Y2 := Ypoints[j+1];
                    generate (X1,Y1,X2,Y2,level);
                end;
            end
            else
            begin
                for k:=0 to generator_size - 1 do
                begin
                    Line(Round(Xpoints[k]+320),Round(175 -
                        Ypoints[k]*0.729)
                        ,Round(Xpoints[k+1]+320),Round(175
                        - Ypoints[k+1]* 0.729));
                end;
            end;    end;

begin

    generator_size := 8;
    init_size := 6;
    initiator_x1[0] := -75;
    initiator_x1[1] := 75;
    initiator_x1[2] := 150;
    initiator_x1[3] := 75;
    initiator_x1[4] := -75;
    initiator_x1[5] := -150;
    initiator_x2[0] := 75;
    initiator_x2[1] := 150;
    initiator_x2[2] := 75;
    initiator_x2[3] := -75;
    initiator_x2[4] := -150;
    initiator_x2[5] := -75;
    initiator_y1[0] := 115;
```

```
initiator_y1[1] := 115;
initiator_y1[2] := 0;
initiator_y1[3] := -115;
initiator_y1[4] := -115;
initiator_y1[5] := 0;
initiator_y2[0] := 115;
initiator_y2[1] := 0;
initiator_y2[2] := -115;
initiator_y2[3] := -115;
initiator_y2[4] := 0;
initiator_y2[5] := 115;
write('Enter level (1 - 8): ');
readln(level);
if level < 1 then
    level := 1;
GraphDriver := 4;
GraphMode := EGAHi;
InitGraph(graphDriver,GraphMode,'');
SetLineStyle(0,$FFFF,1);
SetColor(15);
for i:=0 to init_size - 1 do
begin
    generate(initiator_x1[i], initiator_y1[i],
        initiator_x2[i], initiator_y2[i], level);
end;
ch := ReadKey;
end.
```

Complicated Generators

Take a look at the generator shown in Figure 6-25. This generator was discovered by Mandelbrot. It is based upon a grid of equilateral triangles. If the generator consisted of line segments connecting points 0, 1, 2, 3, 4, and 11, it would be rather simple. However, a smaller replica of this simple generator has been inserted between points 4 and 9 and then two regular line segments added to complete the generator. Because two different line segment lengths are used, we must use the expression:

$$r_m^D = 1 \qquad \text{(Equation 6-16)}$$

to determine the fractal dimension (See Chapter 2). First, we need to observe that r for the regular sized segments is $1/3$. For the smaller segments, we can use simple trigonometry to ascertain that r is 0.186339. Thus, we have:

$$6(.3333)^D + 5(.186339)^D = 1 \qquad \text{(Equation 6-17)}$$

which gives a fractal dimension of:

$$D = 1.8575 \qquad \text{(Equation 6-18)}$$

We can easily handle the change in segment length by simply recalculating the *turtle_r* (length of turtle step) at the appropriate place in the generate procedure.

Figure 6-25. Generator and Second Level for Complex Generator Curve

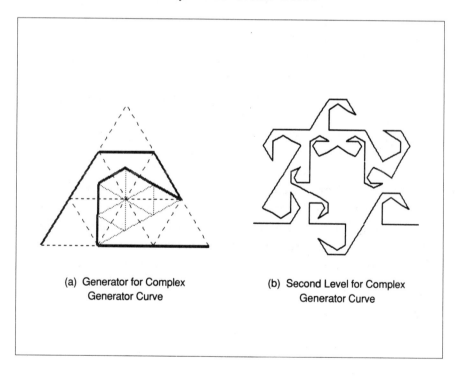

(a) Generator for Complex
Generator Curve

(b) Second Level for Complex
Generator Curve

Now, however, look at Figure 6-26(b), which shows the curve for the second level. In order to make sure the curve is not self-overlapping or self-intersecting, we have to take some considerable liberties with how we use the generator to replace each segment of the previous level. There are four variations of the generator: one is to the right of the original line segment, one is to the left, a third is to the right but with the generator reversed, and the fourth is to the left with the generator reversed. Unfortunately, these are rather arbitrary, and unfortunately, a different set is needed at each level (at least for the first few levels).

We have to thank Mandelbrot for discovering the proper variation to use at each position in the first few levels; it gives us a starting place from which we can branch off into our own investigations. Our software handles the problem by adding two new parameters, sign and type. In using the turtle graphics to create our generator, we multiply every angle by *sign*, which starts out the program with a value of 1. As we enter the generator function, we take action based upon the value of *type*. If *type* is zero, nothing is changed. The parameter *sign* retains its original value and the original generator function is produced on the same side as the previous one. If *type* is 1, the sign of *sign* is reversed, causing all of the turn angles to be reversed, so that the generator appears on the opposite side of the line segment from the previous one. If type is 2, we make the beginning line segment coordinates the end ones and visa-versa, so that the generator is drawn backwards. We also need to reverse all of the signs for this reverse generator to appear on the same side of the line segment as the previous generator. Finally, for a *type* of 3, we reverse coordinates only so that the generator is both reversed and moved to the opposite side.

Figure 6-26. Complex Generator Curves for levels 2 to 4

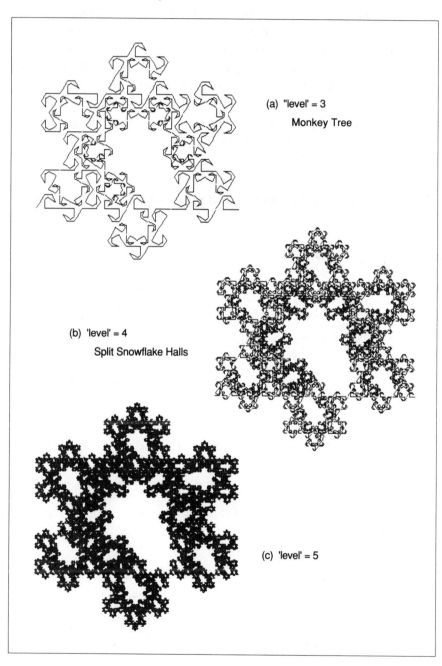

(a) "level' = 3

Monkey Tree

(b) 'level' = 4

Split Snowflake Halls

(c) 'level' = 5

As we enter the recursion process for each level, we have to define what the type is to be for every line segment that is to be replaced. This is a somewhat lengthy process, even using Turbo Pascal's case statement. Fortunately, in this case we only had to define two levels. The net result is shown in Figure 6-26. Mandelbrot calls the third level curve a "Monkey's Tree," and the fourth level curve "Split Snowflake Halls." The fourth level curve is not quite like Mandelbrot's version, because we did not define the *type* parameter to match his for every line segment. Figure 6-27 is the listing of the program to generate these curves.

Figure 6-27. Program to Generate Complex Generator Curves

```
program snowhall;

uses CRT,Graph,Fractal;

var

    graphDriver,GraphMode,i,generator_size,level,init_size:
        integer;
    set_type: integer;
    sign: real;
    Xpoints, Ypoints: array[0..24] of real;
    initiator_x1,initiator_x2,initiator_y1,initiator_y2:
        array[0..9] of real;
    ch: char;


procedure generate (X1: real; Y1: real; X2: real; Y2: real;
    level: integer; gen_type: integer; sign: real);
    var
        j,k,set_type: integer;
        a,b,temp: real;
        Xpoints, Ypoints: array[0..24] of real;

    begin
        case gen_type of
            1: sign := -sign;
            2: begin
                sign := -sign;
                temp := X1;
                X1 := X2;
                X2 := temp;
                temp := Y1;
                Y1 := Y2;
```

```
      Y2 := temp;
    end;
  3: begin
      temp := X1;
      X1 := X2;
      X2 := temp;
      temp := Y1;
      Y1 := Y2;
      Y2 := temp;
      end;
end;
dec(level);
turtle_r := (sqrt((X2 - X1)*(X2 - X1) + (Y2 -
    Y1)*(Y2 - Y1)))/3.0;
Xpoints[0] := X1;
Ypoints[0] := Y1;
Xpoints[11] := X2;
Ypoints[11] := Y2;
turtle_theta := point(X1,Y1,X2,Y2);
turn(60*sign);
turtle_x := X1;
turtle_y := Y1;
step;
Xpoints[1] := turtle_x;
Ypoints[1] := turtle_y;
step;
Xpoints[2] := turtle_x;
Ypoints[2] := turtle_y;
turn(-60*sign);
step;
Xpoints[3] := turtle_x;
Ypoints[3] := turtle_y;
turn(-60*sign);
step;
Xpoints[4] := turtle_x;
Ypoints[4] := turtle_y;
turn(-120*sign);
step;
turn(60*sign);
step;
Xpoints[9] := turtle_x;
Ypoints[9] := turtle_y;
turn(120*sign);
step;
Xpoints[10] := turtle_x;
Ypoints[10] := turtle_y;
turtle_r := (sqrt((Xpoints[9] -
    Xpoints[4])*(Xpoints[9] -
    Xpoints[4]) + (Ypoints[9] -
    Ypoints[4])*(Ypoints[9] -
    Ypoints[4])))/3.0;
```

```
turtle_theta := point(Xpoints[4],
    Ypoints[4],Xpoints[9],Ypoints[9]);
turn(-60*sign);
turtle_x := Xpoints[4];
turtle_y := Ypoints[4];
step;
Xpoints[5] := turtle_x;
Ypoints[5] := turtle_y;
step;
Xpoints[6] := turtle_x;
Ypoints[6] := turtle_y;
turn(60*sign);
step;
Xpoints[7] := turtle_x;
Ypoints[7] := turtle_y;
turn(60*sign);
step;
Xpoints[8] := turtle_x;
Ypoints[8] := turtle_y;
if level = 0 then
begin
    for k:=0 to generator_size - 1 do
    begin
        Line(Round(Xpoints[k]+320),Round(175 -
            Ypoints[k]*0.729)
            ,Round(Xpoints[k+1]+320),Round(175
            - Ypoints[k+1]* 0.729));
    end;
end
else
begin
    for j:=0 to generator_size - 1 do
    begin
        if level = 1 then
            begin
                case j of
                    2,8,10: set_type := 0;
                    0,5:    set_type := 1;
                    1,3,4:  set_type := 2;
                    6,7,9:  set_type := 3;
                end;
            end;
        if level > 1 then
        begin
            case j of
                2,8,10: set_type := 0;
                0:      set_type := 1;
                1,3,4:  set_type := 2;
                5,6,7,9:  set_type := 3;
            end;
        end;
    end;
```

```
                X1 := Xpoints[j];
                X2 := Xpoints[j+1];
                Y1 := Ypoints[j];
                Y2 := Ypoints[j+1];
                generate (X1,Y1,X2,Y2,level,
                    set_type,sign);
            end;
        end;
    end;

begin
    sign := 1;
    set_type := 0;
    generator_size := 11;
    init_size := 1;
    initiator_x1[0] := -150;
    initiator_x2[0] := 150;
    initiator_y1[0] := -75;
    initiator_y2[0] := -75;
    write('Enter level (1 - 8): ');
    readln(level);
    if level < 1 then
        level := 1;
    GraphDriver := 4;
    GraphMode := EGAHi;
    InitGraph(graphDriver,GraphMode,'');
    SetLineStyle(0,$FFFF,1);      SetColor(15);
    for i:=0 to init_size - 1 do
    begin
        generate(initiator_x1[i], initiator_y1[i],
            initiator_x2[i], initiator_y2[i], level,set_type,sign);
    end;
    ch := ReadKey;
end.
```

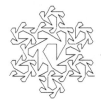

7

Peano Curves

Chapter 6 described a number of curves which were characterized by self-similarity, no self-intersection, and no self-overlapping. They had fractal dimensions greater than 1 and less than 2, which implies that no matter how many times the recursion process was applied, the curves would never completely fill the plane. In this chapter, we will consider curves whose fractal dimension, D, is 2. They are called Peano curves, because the first of the family, which will be described in the next section, was discovered by Giuseppe Peano in 1900. The fractal dimension of 2 has two implications. First, the curves must completely fill the plane. Second, the curves must be self-intersecting—if they fill the plane, there must be an infinity of points at which each curve intersects itself.

The Original Peano Curve

Figure 7-1 shows the generator for the original Peano curve. The initiator is simply a horizontal straight line. Unfortunately, because of all of the self-intersections, it is almost impossible to determine the way in which the Peano curve is drawn, even if arrows are added to the diagram in an attempt to show the flow. As you look at the diagram, first a step is made up, then one to the left, then another up, then one to the right, then one down, then one to the right, then one up, then one to the left, and finally one up. Figure 7-2 shows the

Peano curves for levels of 2, 3, and 4. The way in which the generator is drawn can be best understood by looking at the turtle graphics part of the listing for the generator function, which is included in Figure 7-3, the listing of the program to draw this curve. The generator can be seen to consist of nine line segments ($N = 9$), each of which has a length of 1/3 of the original line ($r = 1/3$). Thus, the fractal dimension is:

```
D = log 9 / log 3 = 2          (Equation 7-1)
```

Figure 7-1. Generator for Original Peano Curve

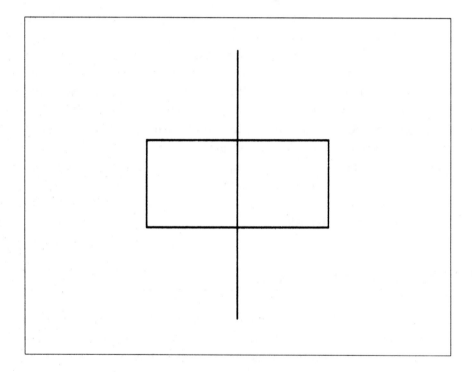

Figure 7-2. Original Peano Curves for levels 2 to 4

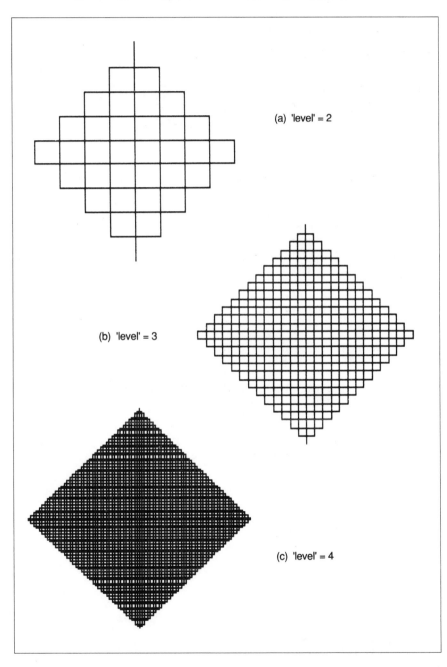

(a) 'level' = 2

(b) 'level' = 3

(c) 'level' = 4

Figure 7-3. Generator for Original Peano Curve

```
program peanoor;

uses CRT,Graph,Fractal;

var
    graphDriver,GraphMode,i,generator_size,level,init_size: integer;
        Xpoints, Ypoints: array[0..24] of real;
    initiator_x1,initiator_x2,initiator_y1,initiator_y2: array[0..9]
        of real;
    ch: char;

procedure step;

    begin
        turtle_x := turtle_x +
turtle_r*cos(turtle_theta*0.017453292);
        turtle_y := turtle_y +
turtle_r*sin(turtle_theta*0.017453292);
    end;


procedure turn(angle: real);

    begin
        turtle_theta := turtle_theta + angle;
    end;

procedure generate (X1: real; Y1: real; X2: real; Y2: real;
    level: integer);
    var
        j,k: integer;
        a, b: real;
        Xpoints, Ypoints: array[0..24] of real;

    begin

        dec(level);
        turtle_r := (sqrt((X2 - X1)*(X2 - X1) + (Y2 - Y1)*(Y2 -
            Y1)))/3.0;
        Xpoints[0] := X1;
        Ypoints[0] := Y1;
        Xpoints[9] := X2;
        Ypoints[9] := Y2;
        turtle_theta := point(X1,Y1,X2,Y2);
        turtle_x := X1;
        turtle_y := Y1;
        step;
        Xpoints[1] := turtle_x;
```

```
      Ypoints[1] := turtle_y;
      turn(90);
      step;
      Xpoints[2] := turtle_x;
      Ypoints[2] := turtle_y;
      turn(-90);
      step;
      Xpoints[3] := turtle_x;
      Ypoints[3] := turtle_y;
      turn(-90);
      step;
      Xpoints[4] := turtle_x;
      Ypoints[4] := turtle_y;
      turn(-90);
      step;
      Xpoints[5] := turtle_x;
      Ypoints[5] := turtle_y;
      turn(90);
      step;
      Xpoints[6] := turtle_x;
      Ypoints[6] := turtle_y;
      turn(90);
      step;
      Xpoints[7] := turtle_x;
      Ypoints[7] := turtle_y;
      turn(90);
      step;
      Xpoints[8] := turtle_x;
      Ypoints[8] := turtle_y;
      if level > 0 then
      begin
          for j:=0 to generator_size - 1 do
          begin
              X1 := Xpoints[j];
                X2 := Xpoints[j+1];
                Y1 := Ypoints[j];
                Y2 := Ypoints[j+1];
              generate (X1,Y1,X2,Y2,level);
          end;
      end
      else
      begin
          for k:=0 to generator_size - 1 do
          begin
              Line(Round(Xpoints[k]+320),Round(175 -
                  Ypoints[k]*0.729)
                  ,Round(Xpoints[k+1]+320),Round(175 -
                  Ypoints[k+1]* 0.729));
          end;
      end;
end;
```

```
begin
    generator_size := 9;
    init_size := 1;
    initiator_x1[0] := 0;
    initiator_x2[0] := 0;
    initiator_y1[0] := -100;
    initiator_y2[0] := 100;
    write('Enter level (1 - 8): ');
    readln(level);
    if level < 1 then
        level := 1;
    GraphDriver := 4;
    GraphMode := EGAHi;
    InitGraph(graphDriver,GraphMode,'');
    SetLineStyle(0,$FFFF,1);
    SetColor(15);
    for i:=0 to init_size - 1 do
    begin
        generate(initiator_x1[i], initiator_y1[i],
            initiator_x2[i],
            initiator_y2[i], level);
    end;
    ch := ReadKey;
end.
```

The Peano curves are generated by the same generic program shown in Figure 6-6. Figure 7-3 is the complete program to draw the Peano curve. The similarities of this to the generic program are obvious upon comparison.

Modified Peano Curve

Were it not for the self-intersections of the generator for the original Peano curve, it would be a lot easier to trace the curve and see how it is drawn. Thus, a modification of the Peano curve has been developed which rounds off the corners to avoid self-intersection. The resulting generator is shown in Figure 7-4. It must be noted, however, that this generator can only be used at the lowest level, just before actual curve drawing; if it is used at higher levels, on recursion the program tries to substitute the generator for each diagonal segment that rounds off a corner as well as for the regular line segments. Therefore the generator for the original Peano curve is used at the higher levels. The curve is mathematically interesting because it is not really quite a Peano curve.

Because the generator used in the final recursion is a little shorter in length than that of the original Peano curve, the fractal dimension, D, is slightly less than 2. As the number of recursions increases, the fractal dimension changes; as the number of recursions approaches infinity, the fractal dimension approaches 2 as a limit.

Figure 7-4. Generator for Modified Peano Curve

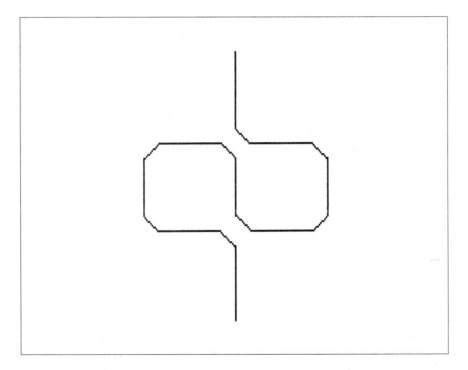

Figure 7-5 shows the resulting modified Peano curves for levels of 2 and 3. To generate these curves, we use the program listed in Figure 7-6. The generator function for all levels above one is the same as for the original Peano curve. For level one, a different generator is used.

Instead of defining a turtle step (*turtle_r*) as one-third of the original line segment, it is defined as one-eighteenth. The basic generator is then written to have the turtle traverse the same path as the original Peano curve generator, which is done by using the same turn angles, but taking six steps for each one that was taken by the original generator. However, the points that are saved for the coordinate array are different.

Figure 7-5. Modified Peano Curves for Levels 2 and 3

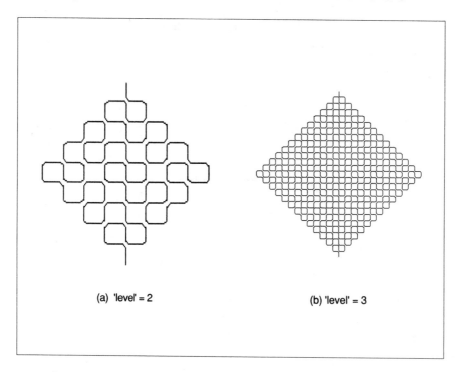

(a) 'level' = 2   (b) 'level' = 3

After saving the first set of coordinates, we next save the location after the fifth step. The next location to be saved is at the end of the first step after the first corner is turned. The remaining locations are after the fifth step of each line segment and after the first step of the next line segment, except that the fifth step of the very last line segment is not saved. The result, when the lines are drawn, is that a diagonal line connects points one-sixth of the distance on each line segment that would normally meet at the corner.

Figure 7-6. Program to Generate Modified Peano Curves

```
program peanomod;

uses CRT,Graph,Fractal;

var
    graphDriver,GraphMode,i,generator_size,level,init_size:
        integer;
    Xptemp,Yptemp: real;
    Xpoints, Ypoints: array[0..24] of real;
    initiator_x1,initiator_x2,initiator_y1,initiator_y2:
        array[0..9] of real;
    ch: char;

procedure generate (X1: real; Y1: real; X2: real; Y2: real;
    level: integer);
    var
        j,k: integer;
        a, b: real;
        Xpoints, Ypoints: array[0..24] of real;

    begin

        dec(level);
        Xpoints[0] := X1;
        Ypoints[0] := Y1;
        turtle_theta := point(X1,Y1,X2,Y2);
        turtle_x := X1;
        turtle_y := Y1;
        if level <> 0 then
        begin
            turtle_r := (sqrt((X2 - X1)*(X2 - X1) + (Y2 -
                Y1)* (Y2 - Y1)))/3.0;
            Xpoints[9] := X2;
            Ypoints[9] := Y2;
            step;
            Xpoints[1] := turtle_x;
            Ypoints[1] := turtle_y;
            turn(90);                    step;
            Xpoints[2] := turtle_x;
            Ypoints[2] := turtle_y;
            turn(-90);
            step;
```

```
        Xpoints[3] := turtle_x;
        Ypoints[3] := turtle_y;
        turn(-90);
        step;
        Xpoints[4] := turtle_x;
        Ypoints[4] := turtle_y;
        turn(-90);
        step;
        Xpoints[5] := turtle_x;
        Ypoints[5] := turtle_y;
        turn(90);
        step;
        Xpoints[6] := turtle_x;
        Ypoints[6] := turtle_y;
        turn(90);
        step;
        Xpoints[7] := turtle_x;
        Ypoints[7] := turtle_y;
        turn(90);
        step;
        Xpoints[8] := turtle_x;
        Ypoints[8] := turtle_y;
        for j:=0 to 8 do
        begin
            X1 := Xpoints[j];
            X2 := Xpoints[j+1];
            Y1 := Ypoints[j];
            Y2 := Ypoints[j+1];
            generate (X1,Y1,X2,Y2,level);
        end;
end
else
begin
    turtle_r := (sqrt((X2 - X1)*(X2 - X1) + (Y2 -
        Y1)* (Y2 - Y1)))/18.0;
    Xpoints[0] := Xptemp;
    Ypoints[0] := Yptemp;
    Xpoints[19] := X2;
    Ypoints[19] := Y2;
    step;
    Xpoints[1] := turtle_x;
    Ypoints[1] := turtle_y;
    step;
    step;
    step;
    Xpoints[2] := turtle_x;
    Ypoints[2] := turtle_y;
    step;
    turn(90);
    step;
    Xpoints[3] := turtle_x;
```

```
Ypoints[3] := turtle_y;
step;
step;
step;
step;
Xpoints[4] := turtle_x;
Ypoints[4] := turtle_y;
step;
turn(-90);
step;
Xpoints[5] := turtle_x;
Ypoints[5] := turtle_y;
step;
step;
step;
step;
Xpoints[6] := turtle_x;
Ypoints[6] := turtle_y;
step;
turn(-90);
step;
Xpoints[7] := turtle_x;
Ypoints[7] := turtle_y;
step;
step;
step;
step;
Xpoints[8] := turtle_x;
Ypoints[8] := turtle_y;
step;
turn(-90);
step;
Xpoints[9] := turtle_x;
Ypoints[9] := turtle_y;
step;
step;
step;
step;
Xpoints[10] := turtle_x;
Ypoints[10] := turtle_y;
step;
turn(90);
step;
Xpoints[11] := turtle_x;
Ypoints[11] := turtle_y;
step;
step;
step;
step;
Xpoints[12] := turtle_x;
Ypoints[12] := turtle_y;
```

```
                step;
                turn(90);
                step;
                Xpoints[13] := turtle_x;
                Ypoints[13] := turtle_y;
                step;
                step;
                step;
                step;
                Xpoints[14] := turtle_x;
                Ypoints[14] := turtle_y;
                step;
                turn(90);
                step;
                Xpoints[15] := turtle_x;
                Ypoints[15] := turtle_y;
                step;
                step;
                step;
                step;
                Xpoints[16] := turtle_x;
                Ypoints[16] := turtle_y;
                step;
                turn(-90);
                step;
                Xpoints[17] := turtle_x;
                Ypoints[17] := turtle_y;
                step;
                step;
                step;
                step;
                Xpoints[18] := turtle_x;
                Ypoints[18] := turtle_y;
                Xptemp := Xpoints[18];
                Yptemp := Ypoints[18];
                for k:=0 to generator_size - 2 do
                begin
                    Line(Round(Xpoints[k]+320),Round(175 -
                        Ypoints[k]*0.729)
                        ,Round(Xpoints[k+1]+320),Round(175
                        - Ypoints[k+1]* 0.729));
                end;
            end;
        end;

begin

    generator_size := 19;
    init_size := 1;
    initiator_x1[0] := 0;
    initiator_x2[0] := 0;
```

```
    initiator_y1[0] := -200;
    initiator_y2[0] := 200;
    write('Enter level (1 - 8): ');
    readln(level);
    if level < 1 then
        level := 1;
    GraphDriver := 4;
    GraphMode := EGAHi;
    InitGraph(graphDriver,GraphMode,'');
    SetLineStyle(0,$FFFF,1);
    SetColor(15);
    Xptemp := initiator_x1[0];
    Yptemp := initiator_y1[0];
    for i:=0 to init_size - 1 do
    begin
        generate(initiator_x1[i], initiator_y1[i],
            initiator_x2[i], initiator_y2[i], level);
    end;
    ch := ReadKey;
end.
```

Cesaro Triangle Curve

Figure 7-7(a) shows the very simple generator that will be used for the next few curves. The initiator in each case will be a horizontal straight line. The generator consists of two sides of a right isosceles triangle. Consequently, $N=2$ and $r = 1/2$. Therefore, the fractal dimension is:

$$D = \log 2 \; / \; \log \left(\sqrt{2} \right) = 2 \qquad \text{Equation 7-2}$$

Depending upon the conditions which determine whether this generator is placed to the left or right of each line segment it replaces, many totally different curves can be produced. The first of these to be considered is the Cesaro triangle, discovered by Ernest Cesaro in 1905. Figure 7-7(b) shows the first level of this curve. For any level of construction for this curve, the generator is placed to the right for each line segment at the top level, at the left for each line segment of the next lower level, at the right for each line segment of the next lower level, and so forth.

To do this in our program, we multiply the 90-degree turn angle in the generator by one of an array of *sign*. This parameter is set up at the beginning of the program to be +1 for the top level, and alternate in sign for each succeeding lower level. Figure 7-8 shows the resulting Cesaro triangles for levels 2, 4, 8, and 12. The program to generate this curve is listed in Figure 7-9.

Figure 7-7. Generator and First Level for Cesaro Curve

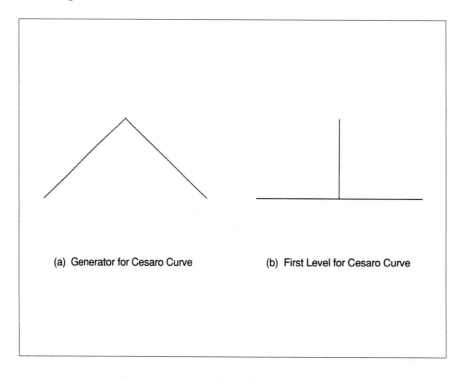

(a) Generator for Cesaro Curve   (b) First Level for Cesaro Curve

Figure 7-8. Cesaro Triangle Curves for Levels 2-12.

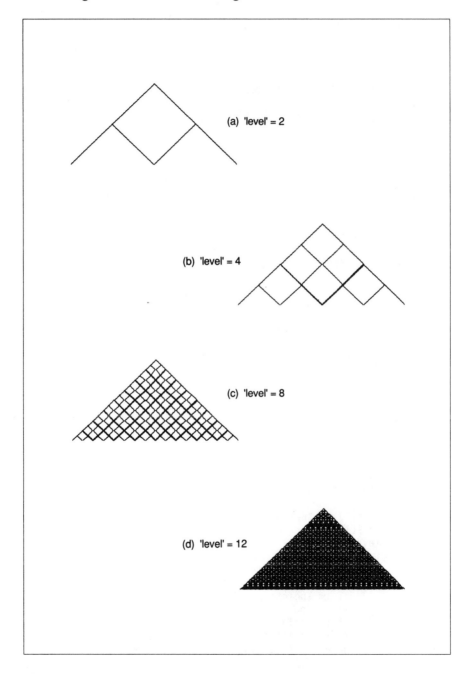

(a) 'level' = 2

(b) 'level' = 4

(c) 'level' = 8

(d) 'level' = 12

Figure 7-9. Program to Generate Cesaro Triangle Curves

```
program cesaro;

uses CRT,Graph,Fractal;

var
      sign1,graphDriver,GraphMode,i,generator_size,level,
      init_size: integer;
   Xpoints, Ypoints: array[0..24] of real;
   initiator_x1,initiator_x2,initiator_y1,initiator_y2:
      array[0..9] of real;
   ch: char;
   sign: array[0..15] of integer;

procedure generate (X1: real; Y1: real; X2: real; Y2: real;
   level: integer);
   var
      j,k: integer;
      a, b: real;
      Xpoints, Ypoints: array[0..24] of real;

   begin

      dec(level);
      turtle_r := sqrt((X2 - X1)*(X2 - X1) + (Y2 -
         Y1)*(Y2 - Y1))/2.0;
      Xpoints[0] := X1;
      Ypoints[0] := Y1;
      Xpoints[2] := X2;
      Ypoints[2] := Y2;
      turtle_theta := point(X1,Y1,X2,Y2);
      turtle_x := X1;
      turtle_y := Y1;
      step;
      Xpoints[3] := turtle_x;
      Ypoints[3] := turtle_y;
      turn(90*sign[level]);
      step;
      Xpoints[1] := turtle_x;
      Ypoints[1] := turtle_y;
      if level > 0 then
      begin
          for j:=0 to generator_size - 2 do
          begin
              X1 := Xpoints[j];
              X2 := Xpoints[j+1];
              Y1 := Ypoints[j];
              Y2 := Ypoints[j+1];
              generate (X1,Y1,X2,Y2,level);
```

```
                    end;
                end
                else
                begin
                    Line(Round(Xpoints[0]+320),Round(175 -
                        Ypoints[0]*0.729)
                        ,Round(Xpoints[2]+320),Round(175 -
                        Ypoints[2]* 0.729));
                    Line(Round(Xpoints[1]+320),Round(175 -
                        Ypoints[1]*0.729)
                        ,Round(Xpoints[3]+320),Round(175 -
                        Ypoints[3]* 0.729));
                end;
        end;

begin

    generator_size := 3;
    init_size := 1;
    initiator_x1[0] := -150;
    initiator_x2[0] := 150;
    initiator_y1[0] := 0;
    initiator_y2[0] := 0;
    write('Enter level (1 - 16): ');
    readln(level);
    if level < 1 then
        level := 1;
    GraphDriver := 4;
    GraphMode := EGAHi;
    InitGraph(graphDriver,GraphMode,'');
    SetLineStyle(0,$FFFF,1);
    SetColor(15);
    sign1 := -1;
    for i:=level downto 0  do
    begin
        sign[i] := sign1;
        sign1 := -sign1;
    end;
    for i:=0 to init_size - 1 do
    begin
        generate(initiator_x1[i], initiator_y1[i],
            initiator_x2[i], initiator_y2[i], level);
    end;
    ch := ReadKey;
end.
```

Modified Cesaro Triangle Curve

The Cesaro triangle curve described above is a little hard to trace because the line going out at right angles from the center of the original line segment actually retraces itself, but this is not possible to observe in the drawings. A modification of the Cesaro curve is possible by changing the angle of the generator from 90 degrees to 85 degrees for the lowest level before drawing occurs. As with the modified Peano curve, this results in a curve whose fractal dimension is not quite 2, but which approaches 2 as a limit when the number of recursions approaches infinity. Figure 7-10 shows the first level for the modified Cesaro Triangle curve. Figure 7-11 shows the resulting curves for levels 2, 4, 8, and 12. The program to generate this curve is that listed in Figure 7-9 with the generator function replaced by that listed in Figure 7-12. We've chosen a slightly different approach here than was used for the modified Peano curve. We generate the three points that are used in the unmodified generator and use them for each step in the recursion process. We also generate two additional points to locate the apex of the 85-degree angle of the two triangles for the first level and use them in drawing the actual curve.

Figure 7-10. First Level for Modified Cesaro Curve

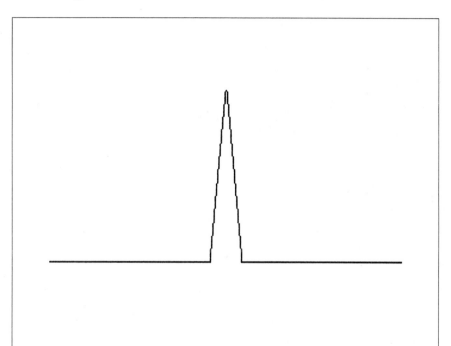

Figure 7-11. Modified Cesaro Triangle Curves for Levels 2-12.

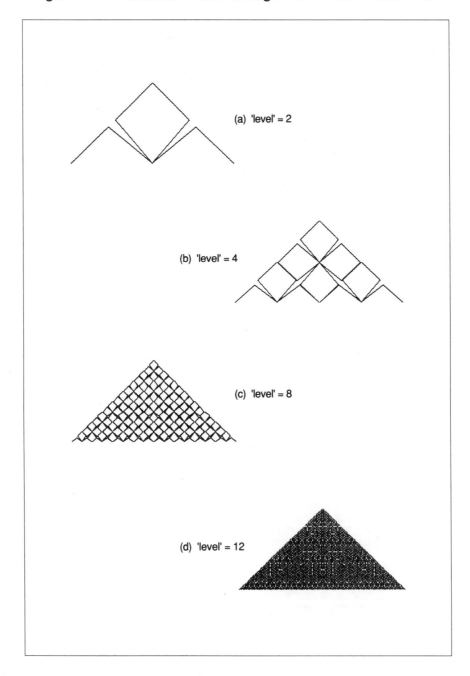

(a) 'level' = 2

(b) 'level' = 4

(c) 'level' = 8

(d) 'level' = 12

Figure 7-12. Generator for Modified Cesaro Curves

```
program cesaro;

uses CRT,Graph,Fractal;

var
    sign1,graphDriver,GraphMode,i,generator_size,level,
        init_size: integer;
    Xpoints, Ypoints: array[0..24] of real;
    initiator_x1,initiator_x2,initiator_y1,initiator_y2:
        array[0..9] of real;
    ch: char;
    sign: array[0..15] of integer;

procedure generate (X1: real; Y1: real; X2: real; Y2: real;
    level: integer);
    var
        j,k: integer;
        a, b: real;
        Xpoints, Ypoints: array[0..24] of real;

    begin

        dec(level);
        turtle_r := sqrt((X2 - X1)*(X2 - X1) + (Y2 - Y1)*(Y2
            - Y1))/2.0;
        Xpoints[0] := X1;
        Ypoints[0] := Y1;
        Xpoints[2] := X2;
        Ypoints[2] := Y2;
        turtle_theta := point(X1,Y1,X2,Y2);
        turtle_x := X1;
        turtle_y := Y1;
        step;
        Xpoints[3] := turtle_x;
        Ypoints[3] := turtle_y;
        turn(90*sign[level]);
        step;
        Xpoints[1] := turtle_x;
        Ypoints[1] := turtle_y;
        if level > 0 then
        begin
            for j:=0 to generator_size - 2 do
```

```
            begin
                X1 := Xpoints[j];
                  X2 := Xpoints[j+1];
                  Y1 := Ypoints[j];
                  Y2 := Ypoints[j+1];
                generate (X1,Y1,X2,Y2,level);
            end;
        end
        else
        begin
            Line(Round(Xpoints[0]+320),Round(175 -
            Ypoints[0]*0.729)
                ,Round(Xpoints[2]+320),Round(175 -
                Ypoints[2]* 0.729));
            Line(Round(Xpoints[1]+320),Round(175 -
                Ypoints[1]*0.729)
                ,Round(Xpoints[3]+320),Round(175 -
                Ypoints[3]* 0.729));
        end;
    end;

begin

    generator_size := 3;
    init_size := 1;
    initiator_x1[0] := -150;
    initiator_x2[0] := 150;
    initiator_y1[0] := 0;
    initiator_y2[0] := 0;
    write('Enter level (1 - 16): ');
    readln(level);
    if level < 1 then
        level := 1;
    GraphDriver := 4;
    GraphMode := EGAHi;
    InitGraph(graphDriver,GraphMode,'');
    SetLineStyle(0,$FFFF,1);
    SetColor(15);
    sign1 := -1;
    for i:=level downto 0  do
    begin
        sign[i] := sign1;
        sign1 := -sign1;
```

```
    end;
    for i:=0 to init_size - 1 do
    begin
        generate(initiator_x1[i], initiator_y1[i],
            initiator_x2[i], initiator_y2[i], level);
    end;
    ch := ReadKey;
end.
```

Variation on the Cesaro Curve

Suppose we start with a curve which has the same generator and the same first two levels as the Cesaro curve, but then uses a differing arrangement of placing the generator to the right and left of the original line segment as we go to higher levels. Many different curves can result. One of them is shown for levels 2, 4, 8, and 16 in Figure 7-13. The program that was used to generate these curves is listed in Figure 7-14. This can serve as a basis for your experimentation with various methods of arranging the generator to create a variety of interesting curves.

Figure 7-13. Variation of Cesaro Curves for Levels 2-16.

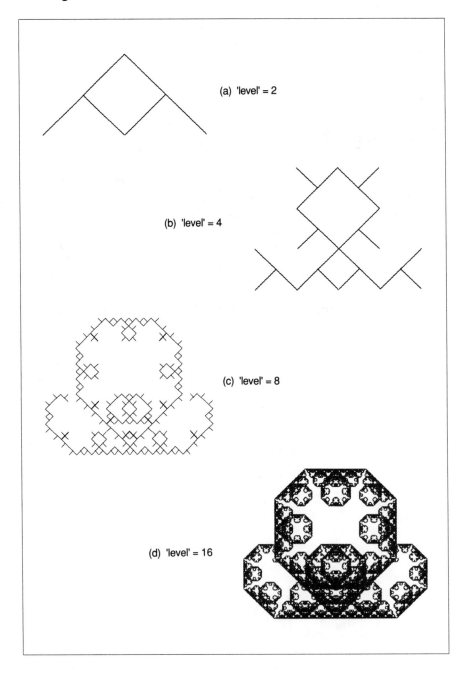

Figure 7-14. Program to Generate Variation of Cesaro Curve

```
program cesaro2;

uses CRT,Graph,Fractal;

var
    sign,graphDriver,GraphMode,i,generator_size,level,
        init_size: integer;
    Xpoints, Ypoints: array[0..24] of real;
    initiator_x1,initiator_x2,initiator_y1,initiator_y2:
        array[0..9] of real;
    ch: char;

procedure generate (X1: real; Y1: real; X2: real; Y2: real;
    level: integer);
    var
        j,k: integer;
        a, b: real;
        Xpoints, Ypoints: array[0..24] of real;

    begin

        dec(level);
        turtle_r := sqrt((X2 - X1)*(X2 - X1) + (Y2 -
            Y1)*(Y2 - Y1))/2.0;
        Xpoints[0] := X1;
        Ypoints[0] := Y1;
        Xpoints[2] := X2;
        Ypoints[2] := Y2;
        turtle_theta := point(X1,Y1,X2,Y2);
        turtle_x := X1;
        turtle_y := Y1;
        step;
        Xpoints[3] := turtle_x;
        Ypoints[3] := turtle_y;
        turn(90*sign);
        step;
        Xpoints[1] := turtle_x;
        Ypoints[1] := turtle_y;
        sign := -1;
        if level > 0 then
        begin
            for j:=0 to generator_size - 2 do
            begin
                X1 := Xpoints[j];
                X2 := Xpoints[j+1];
                Y1 := Ypoints[j];
                Y2 := Ypoints[j+1];
                generate (X1,Y1,X2,Y2,level);
```

```
                end;
            end
            else
            begin
                Line(Round(Xpoints[0]+320),Round(175 -
                    Ypoints[0]*0.729)
                    ,Round(Xpoints[2]+320),Round(175 -
                    Ypoints[2]* 0.729));
                Line(Round(Xpoints[1]+320),Round(175 -
                    Ypoints[1]*0.729)
                    ,Round(Xpoints[3]+320),Round(175 -
                    Ypoints[3]* 0.729));
            end;
        end;

begin

    generator_size := 3;
    init_size := 1;
    initiator_x1[0] := -150;
    initiator_x2[0] := 150;
    initiator_y1[0] := 0;
    initiator_y2[0] := 0;
    write('Enter level (1 - 16): ');
    readln(level);
    if level < 1 then
        level := 1;
    GraphDriver := 4;
    GraphMode := EGAHi;
    InitGraph(graphDriver,GraphMode,'');
    SetLineStyle(0,$FFFF,1);
    SetColor(15);
    sign := 1;
    for i:=0 to init_size - 1 do
    begin
        generate(initiator_x1[i], initiator_y1[i],
            initiator_x2[i], initiator_y2[i], level);
    end;
    ch := ReadKey;
end.
```

Polya Triangle Curve

This curve was discovered by G. Polya. The initiator and generator
are the same as for the Cesaro curve, but the positioning of the generator is different. Figure 7-15 shows the first and second levels of the

curve. As with the Cesaro curve, the position of the first generator alternates from right to left beginning at the top level. For this curve, the position of the generator also alternates with each line segment of a particular level that is replaced. Figure 7-16 shows the resulting curve for levels of 4, 8, and 12. Figure 7-17 lists the program for generating the Polya curves. We use the same technique that was used for the Cesaro curve of having an array of *sign* variables, which are initiated at the beginning of the program. For this curve, we also modify the sign as we pass through the *generate* function. In Chapter 14, we shall discuss the Harter-Heightway dragon curve. Although it is included with the dragons, it is a member of the family of Peano curves discussed in this chapter. It has the same initiator, generator, and first stage as the Polya triangle curve, but then diverges.

Figure 7-15. First Two Levels for Polya Triangle Curve

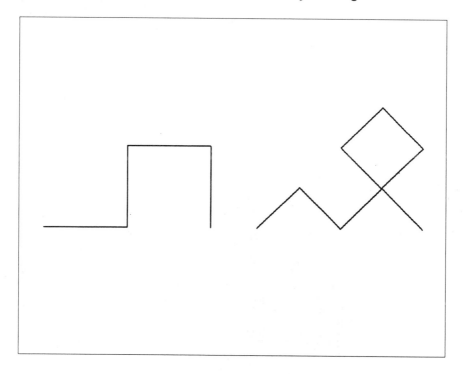

Figure 7-16. Polya Triangle Curves for levels 4 to 12

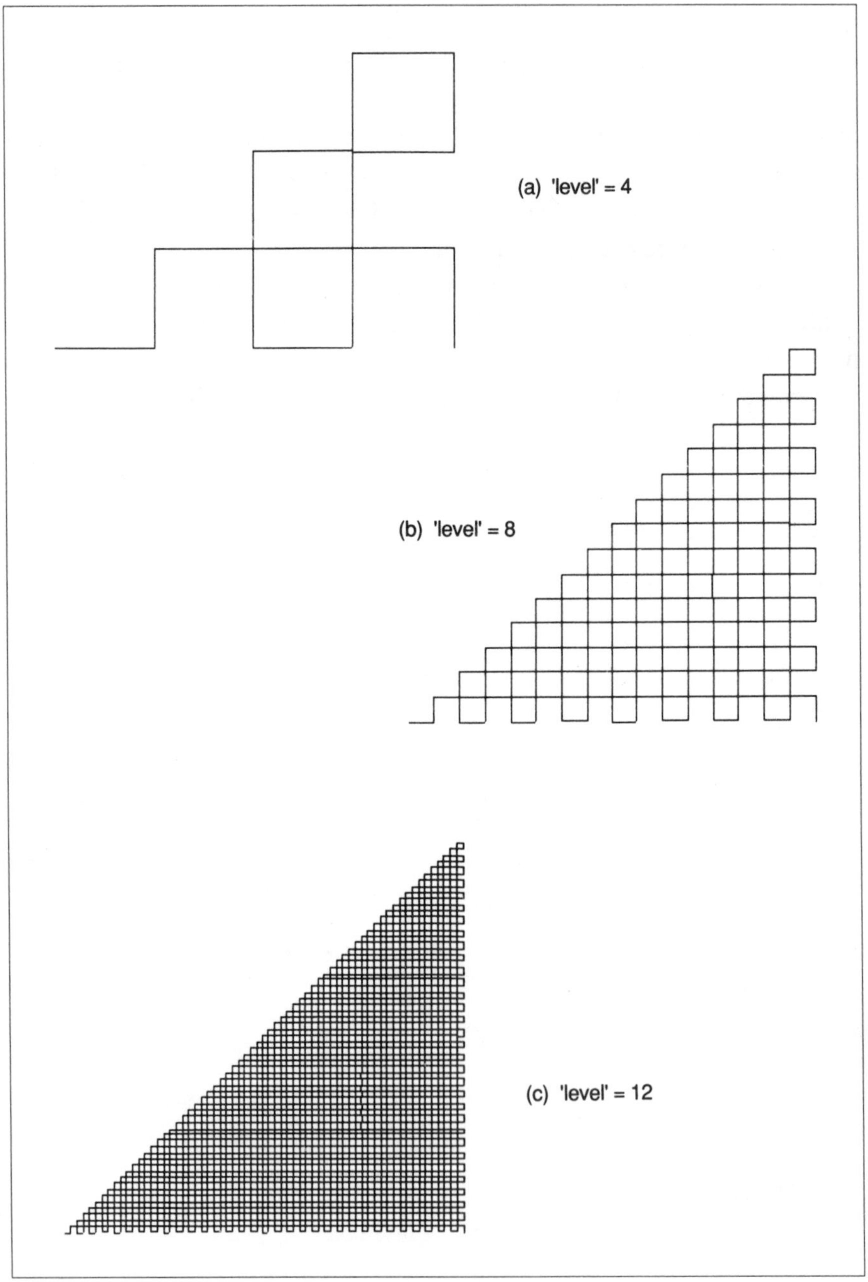

(a) 'level' = 4

(b) 'level' = 8

(c) 'level' = 12

Figure 7-17. Program to Generate Polya Curves

```
program polya;

uses CRT,Graph,Fractal;

var
    sign1,graphDriver,GraphMode,i,generator_size,level,
        init_size: integer;
    Xpoints, Ypoints: array[0..24] of real;
    initiator_x1,initiator_x2,initiator_y1,initiator_y2:
        array[0..9] of real;
    ch: char;
    sign: array[0..16] of integer;

procedure generate (X1: real; Y1: real; X2: real; Y2: real;
    level: integer);
    var
        j,k: integer;
        a, b: real;
        Xpoints, Ypoints: array[0..24] of real;

    begin
        turtle_r := (sqrt((X2 - X1)*(X2 - X1) + (Y2 -
            Y1)*(Y2 - Y1)))/1.41421;
        Xpoints[0] := X1;
        Ypoints[0] := Y1;
        Xpoints[2] := X2;
        Ypoints[2] := Y2;
        turtle_theta := point(X1,Y1,X2,Y2);
        turtle_x := X1;
        turtle_y := Y1;
        turn(sign[level]*45);
        step;
        Xpoints[1] := turtle_x;
        Ypoints[1] := turtle_y;
        dec(level);
        if level > 0 then
        begin
            for j:=0 to generator_size - 2 do
            begin
                X1 := Xpoints[j];
                X2 := Xpoints[j+1];
                Y1 := Ypoints[j];
                Y2 := Ypoints[j+1];
                generate (X1,Y1,X2,Y2,level);
                sign[level] := -sign[level];
            end;
        end
        else
```

```
        begin
            for k:=0 to generator_size - 2 do
                Line(Round(Xpoints[k]+320),Round(175 -
                Ypoints[k]*0.729)
                ,Round(Xpoints[k+1]+320),Round(175 -
                Ypoints[k+1]* 0.729));
        end;
    end;

begin

    sign1 := 1;
    generator_size := 3;
    init_size := 2;
    initiator_x1[0] := -150;
    initiator_x2[0] := 150;
    initiator_y1[0] := -75;
    initiator_y2[0] := -75;
    write('Enter level (1 - 16): ');
    readln(level);
    if level < 1 then
        level := 1;
    GraphDriver := 4;
    GraphMode := EGAHi;
    InitGraph(graphDriver,GraphMode,'');
        SetLineStyle(0,$FFFF,1);
    SetColor(15);
    for i:=level downto 0  do
    begin
        sign[i] := sign1;
        sign1 := -sign1;
    end;
    for i:=0 to init_size - 2 do
    begin
        generate(initiator_x1[i], initiator_y1[i],
            initiator_x2[i], initiator_y2[i], level);
    end;
    ch := ReadKey;
end.
```

The Peano-Gosper Curve

Figure 7-18 shows the generator for the Peano-Gosper curve and its associated grid of equilateral triangles. The geometry of the situation can easily be determined from this figure. There are seven line segments ($N=7$) and the length of each one is:

$$r = 1/ \sqrt{7} \qquad\qquad \text{Equation 7-3}$$

The fractal dimension is

$$D = \log 7 / \log (\sqrt{7}) = 2 \qquad\qquad \text{Equation 7-4}$$

This curve has the interesting characteristic that it just fills the interior of Gosper curve given in Chapter 6. Figure 7-19 shows the curves for levels of 2, 3, and 4. The program is the generic program of Figure 6-6 with a new *generate* procedure. We need to make use of a *generate* procedure which provides for specifying any of the four possible positions of the generator as we did with several of the von Koch curves. The resulting program listing for the Peano-Gosper curve is given in Figure 7-20.

Figure 7-18. Initiator for Peano-Gosper Curve

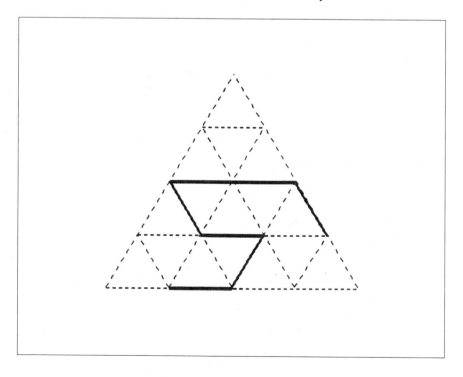

Figure 7-19. Peano-Gosper Curves for levels 2 to 4 Figure

(a) 'level' = 2

(b) 'level' = 3

(c) 'level' = 4

7-20. Program to Generate the Peano-Gosper Curve

```pascal
program pgosper;

uses CRT,Graph,Fractal;

var
    graphDriver,GraphMode,i,generator_size,level,init_size:
        integer;
    gen_type: integer;
    sign: real;
    Xpoints, Ypoints: array[0..24] of real;
    initiator_x1,initiator_x2,initiator_y1,initiator_y2:
        array[0..9] of real;
    ch: char;

procedure generate (X1: real; Y1: real; X2: real; Y2: real;
    level: integer; gen_type: integer);
    var
        j,k,set_type: integer;
        a,b,temp: real;
        Xpoints, Ypoints: array[0..24] of real;

    begin
        case gen_type of
            1: sign := -sign;
            2: begin
                sign := -sign;
                temp := X1;
                X1 := X2;
                X2 := temp;
                temp := Y1;
                Y1 := Y2;
                Y2 := temp;
                end;
            3: begin
                temp := X1;
                X1 := X2;
                X2 := temp;
                temp := Y1;
                Y1 := Y2;
                Y2 := temp;
                end;
        end;
        dec(level);
        turtle_r := (sqrt((X2 - X1)*(X2 - X1) + (Y2 -
            Y1)*(Y2 - Y1)))/ 2.6457513;
        Xpoints[0] := X1;
        Ypoints[0] := Y1;
        Xpoints[7] := X2;
```

```
            Ypoints[7] := Y2;
            turtle_theta := point(X1,Y1,X2,Y2);
            turn(-19*sign);
            turtle_x := X1;
            turtle_y := Y1;
            step;
            Xpoints[1] := turtle_x;
            Ypoints[1] := turtle_y;
            turn(60*sign);
            step;
            Xpoints[2] := turtle_x;
            Ypoints[2] := turtle_y;
            turn(120*sign);
            step;
            Xpoints[3] := turtle_x;
            Ypoints[3] := turtle_y;
            turn(-60*sign);
            step;
            Xpoints[4] := turtle_x;
            Ypoints[4] := turtle_y;
            turn(-120*sign);
            step;
            Xpoints[5] := turtle_x;
            Ypoints[5] := turtle_y;
            step;
            Xpoints[6] := turtle_x;
            Ypoints[6] := turtle_y;
            if level = 0 then
            begin
                for k:=0 to generator_size - 2 do
                begin
Line(Round(Xpoints[k]+320),Round(175 -
                    Ypoints[k]*0.729)
                    ,Round(Xpoints[k+1]+320),Round(175
                    - Ypoints[k+1]* 0.729));
                end;
            end
            else
            begin
                for j:=0 to generator_size - 2 do
                begin
                    case j of
                        0,3,4,5: set_type := 0;
                        2,1,6:   set_type := 3;
                    end;
                X1 := Xpoints[j];
                X2 := Xpoints[j+1];
                Y1 := Ypoints[j];
                Y2 := Ypoints[j+1];
                generate (X1,Y1,X2,Y2,level,set_type);
                end;
```

```
        end;
    end;

begin
    sign := 1;
    gen_type := 3;
    generator_size := 8;
    init_size := 1;
    initiator_x1[0] := -150;
    initiator_x2[0] := 150;
    initiator_y1[0] := 75;
    initiator_y2[0] := 75;
    write('Enter level (1 - 8): ');
    readln(level);
    if level < 1 then
        level := 1;
    GraphDriver := 4;
    GraphMode := EGAHi;
    InitGraph(graphDriver,GraphMode,'');
    SetLineStyle(0,$FFFF,1);
    SetColor(15);
    for i:=0 to init_size - 1 do
    begin
        generate(initiator_x1[i], initiator_y1[i],
            initiator_x2[i], initiator_y2[i], level,gen_type);
    end;
    ch := ReadKey;
end.
```

Peano Seven Segment Snowflake

Figure 7-21 shows the generator and first stage of a Peano seven seg-ment snowflake curve discovered by Mandelbrot. Note the similarity of the generator to that described under the heading *Complicated Generators* in Chapter 6. The only difference is that where the gener-ator of Chapter 6 used a smaller replica of the curve consisting of the first four line segments and then a line to the end to replace the fifth line segment, this curve does not. The result is that the fractal dimen-sion is different. It is:

$$6(1/3)D + (\sqrt{3}/3)D = 1 \qquad \text{Equation 7-5}$$

which gives a fractal dimension of:

$$D = 2 \hspace{4cm} \text{Equation 7-6}$$

Like the complicated generator of Chapter 6, there are four choices of generator position and they must be carefully selected for each level and each line segment to assure that the curve is not self-intersecting or self-overlapping. Figure 7-22 shows the curve for levels of 2, 3, and 4. The program to generate this curve is given in Figure 7-23.

Figure 7-21. First Two Levels for Peano Seven Segment Snowflake Curve

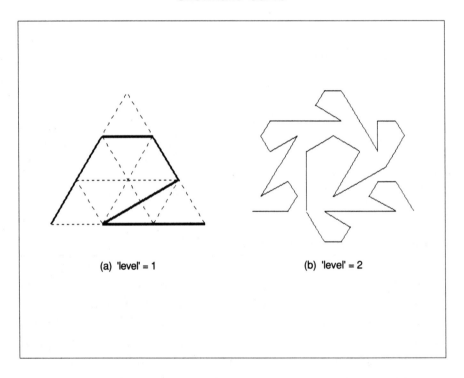

(a) 'level' = 1   (b) 'level' = 2

Figure 7-22. Peano Seven Segment Snowflake Curves for levels 3 to 5

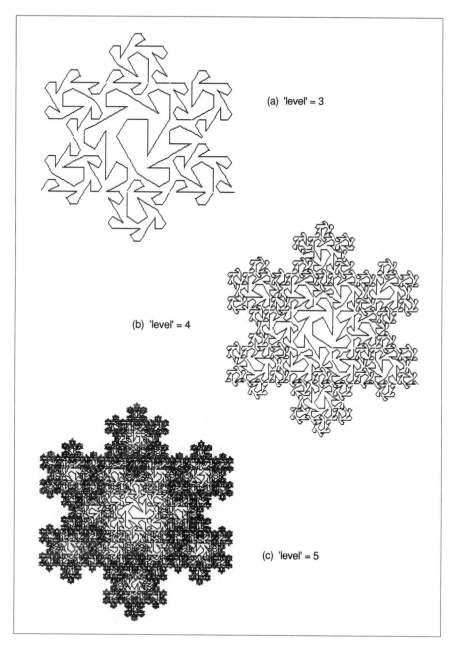

(a) 'level' = 3

(b) 'level' = 4

(c) 'level' = 5

Figure 7-23. Program to Generate Seven Segment Snowflake

```pascal
program snow7;

uses CRT,Graph,Fractal;

var
    graphDriver,GraphMode,i,generator_size,level,init_size:
        integer;
    gen_type,sign: integer;
    Xpoints, Ypoints: array[0..24] of real;
    initiator_x1,initiator_x2,initiator_y1,initiator_y2:
        array[0..9] of real;
    ch: char;

procedure generate (X1: real; Y1: real; X2: real; Y2: real;
    level: integer; gen_type: integer; sign: integer);
    var
        j,k,set_type: integer;
        temp: real;
        Xpoints, Ypoints: array[0..24] of real;

    begin
        case gen_type of
            1: begin
                 sign := -sign;
               end;
            2: begin
                 sign := -sign;
                 temp := X1;
                 X1 := X2;
                 X2 := temp;
                 temp := Y1;
                 Y1 := Y2;
                 Y2 := temp;
               end;
            3: begin
                 temp := X1;
                 X1 := X2;
                 X2 := temp;
                 temp := Y1;
                 Y1 := Y2;
                 Y2 := temp;
               end;
        end;
        dec(level);
        turtle_r := (sqrt((X2 - X1)*(X2 - X1) + (Y2 -
            Y1)*(Y2 - Y1)))/3.0;
        Xpoints[0] := X1;
        Ypoints[0] := Y1;
```

```
Xpoints[7] := X2;
Ypoints[7] := Y2;
turtle_theta := point(X1,Y1,X2,Y2);
turtle_x := X1;
turtle_y := Y1;
turn(60*sign);
step;
Xpoints[1] := turtle_x;
Ypoints[1] := turtle_y;
step;
Xpoints[2] := turtle_x;
Ypoints[2] := turtle_y;
turn(-60*sign);
step;
Xpoints[3] := turtle_x;
Ypoints[3] := turtle_y;
turn(-60*sign);
step;
Xpoints[4] := turtle_x;
Ypoints[4] := turtle_y;
turn(-60*sign);
step;
Xpoints[6] := turtle_x;
Ypoints[6] := turtle_y;
turn(-60*sign);
step;
Xpoints[5] := turtle_x;
Ypoints[5] := turtle_y;
if level = 0 then
begin
    for k:=0 to generator_size - 1 do
    begin
        Line(Round(Xpoints[k]+320),Round(175 -
            Ypoints[k]*0.729)
            ,Round(Xpoints[k+1]+320),Round(175
            - Ypoints[k+1]* 0.729));
    end;
end
else
begin
    for j:=0 to generator_size - 1 do
    begin
        case j of
            0,5:     set_type := 1;
            1,2,3,6: set_type := 2;
            4:       set_type := 3;
        end;
    X1 := Xpoints[j];
    X2 := Xpoints[j+1];
    Y1 := Ypoints[j];
    Y2 := Ypoints[j+1];
```

```
            generate (X1,Y1,X2,Y2,level,set_type,sign);
            end;
        end;
    end;

begin
    DirectVideo := false;
    sign := 1;
    generator_size := 7;
    init_size := 1;
    initiator_x1[0] := -125;
    initiator_x2[0] := 125;
    initiator_y1[0] := 0;
    initiator_y2[0] := 0;
    write('Enter level (1 - 8): ');
    readln(level);
    if level < 1 then
        level := 1;
    GraphDriver := 4;
    GraphMode := EGAHi;
    InitGraph(graphDriver,GraphMode,'');
    SetLineStyle(0,$FFFF,1);
    SetColor(15);
    for i:=0 to init_size - 1 do
    begin
        generate(initiator_x1[i], initiator_y1[i],
            initiator_x2[i], initiator_y2[i], level,0,sign);
    end;
    ch := ReadKey;
end.
```

Peano Thirteen Segment Snowflake

Figure 7-24 shows the generator and first stage of a Peano thirteen segment snowflake curve which was also discovered by Mandelbrot. This generator is obtained by replacing the fifth line segment of the generator in Figure 7-21 with a smaller replica of the entire generator of Figure 7-21. To determine the fractal dimension of this curve, we note that Equation 7-5 applied to the curve of the previous section, and that the length of the line segment being replaced was one. Thus, the fractal dimension is unchanged when this curve is substituted for a line segment, and the fractal dimension of the thirteen segment snowflake is still 2. More generally, we ought to be able to substitute a generator for any line segment of the original generator and still keep the fractal dimension unchanged.

For this curve also, there are four choices of generator position which must be carefully selected for each level and each line segment to assure that the curve is not self-intersecting or self-overlapping. Figure 7-25 shows the curve for levels of 2, 3, and 4. The program to generate the thirteen segment curve is listed in Figure 7-26.

Figure 7-24. First Two Levels for Peano Thirteen Segment Snowflake Curve

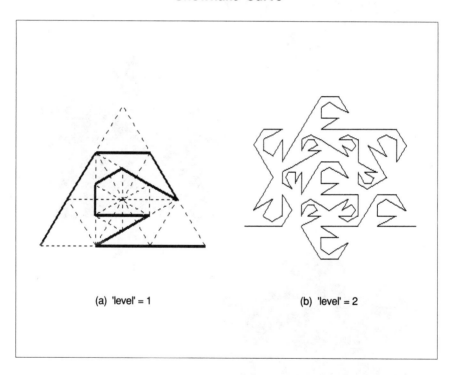

(a) 'level' = 1

(b) 'level' = 2

Figure 7-25. Peano Thirteen Segment Snowflake Curves for levels 3 to 5

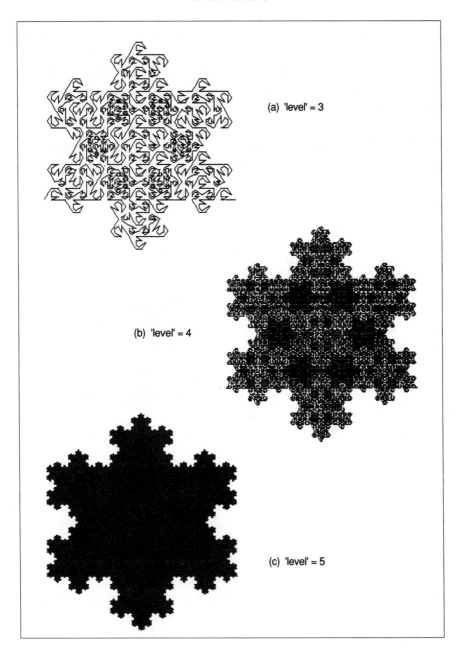

(a) 'level' = 3

(b) 'level' = 4

(c) 'level' = 5

Figure 7-26. Program to Generate Thirteen Segment Snowflake

```
program snow13;

uses CRT,Graph,Fractal;

var
    graphDriver,GraphMode,i,generator_size,level,init_size:
        integer;
    gen_type,sign: integer;
    Xpoints, Ypoints: array[0..24] of real;
    initiator_x1,initiator_x2,initiator_y1,initiator_y2:
        array[0..9] of real;
    ch: char;

procedure generate (X1: real; Y1: real; X2: real; Y2: real;
    level: integer; gen_type: integer; sign: integer);
    var
        j,k,set_type: integer;
        temp: real;
        Xpoints, Ypoints: array[0..24] of real;

    begin
        case gen_type of
            1: begin
                sign := -sign;
                end;
            2: begin
                sign := -sign;
                temp := X1;
                X1 := X2;
                X2 := temp;
                temp := Y1;
                Y1 := Y2;
                Y2 := temp;
                end;
            3: begin
                temp := X1;
                X1 := X2;
                X2 := temp;
                temp := Y1;
                Y1 := Y2;
                Y2 := temp;
                end;
        end;
        dec(level);
        turtle_r := (sqrt((X2 - X1)*(X2 - X1) + (Y2 -
            Y1)*(Y2 - Y1)))/3.0;
        Xpoints[0] := X1;
        Ypoints[0] := Y1;
```

```
Xpoints[13] := X2;
Ypoints[13] := Y2;
turtle_theta := point(X1,Y1,X2,Y2);
turtle_x := X1;
turtle_y := Y1;
turn(60*sign);
step;
Xpoints[1] := turtle_x;
Ypoints[1] := turtle_y;
step;
Xpoints[2] := turtle_x;
Ypoints[2] := turtle_y;
turn(-60*sign);
step;
Xpoints[3] := turtle_x;
Ypoints[3] := turtle_y;
turn(-60*sign);
step;
Xpoints[4] := turtle_x;
Ypoints[4] := turtle_y;
turn(-60*sign);
step;
Xpoints[12] := turtle_x;
Ypoints[12] := turtle_y;
turn(-60*sign);
step;
Xpoints[11] := turtle_x;
Ypoints[11] := turtle_y;
turtle_r := (sqrt((Xpoints[11] -
    Xpoints[4])*(Xpoints[11]
    - Xpoints[4]) + (Ypoints[11] -
    Ypoints[4])*(Ypoints[11] -
    Ypoints[4])))/3.0;
turtle_theta := point(Xpoints[4],
    Ypoints[4],Xpoints[11],Ypoints[11]);
turtle_x := Xpoints[4];
turtle_y := Ypoints[4];
turn(-60*sign);
step;
Xpoints[5] := turtle_x;
Ypoints[5] := turtle_y;
step;
Xpoints[6] := turtle_x;
Ypoints[6] := turtle_y;
turn(60*sign);
step;
Xpoints[7] := turtle_x;
Ypoints[7] := turtle_y;
turn(60*sign);
step;
Xpoints[8] := turtle_x;
```

```
                Ypoints[8] := turtle_y;
                turn(60*sign);
                step;
                Xpoints[10] := turtle_x;
                Ypoints[10] := turtle_y;
                turn(60*sign);
                step;
                Xpoints[9] := turtle_x;
                Ypoints[9] := turtle_y;
                if level = 0 then
                begin
                    for k:=0 to generator_size - 1 do
                    begin
                        Line(Round(Xpoints[k]+320),Round(175 -
                            Ypoints[k]*0.729)
                            ,Round(Xpoints[k+1]+320),Round(175
                            - Ypoints[k+1]* 0.729));
                    end;
                end
                else
                begin
                    for j:=0 to generator_size - 1 do
                    begin
                        case j of
                            1,2,3,4,8,9,12: set_type := 0;
                            0,5,6,7,10,11:  set_type := 1;
                        end;
                    X1 := Xpoints[j];
                    X2 := Xpoints[j+1];
                    Y1 := Ypoints[j];
                    Y2 := Ypoints[j+1];
                    generate (X1,Y1,X2,Y2,level,set_type,sign);
                    end;
                end;
            end;
        end;

begin
    sign := 1;
    generator_size := 13;
    init_size := 1;
    initiator_x1[0] := -125;
    initiator_x2[0] := 125;
    initiator_y1[0] := 0;
    initiator_y2[0] := 0;
    write('Enter level (1 - 8): ');
    readln(level);
    if level < 1 then
        level := 1;
    GraphDriver := 4;
    GraphMode := EGAHi;
    InitGraph(graphDriver,GraphMode,'');
```

```
    SetLineStyle(0,$FFFF,1);
    SetColor(15);
    for i:=0 to init_size - 1 do
    begin
        generate(initiator_x1[i], initiator_y1[i],
            initiator_x2[i], initiator_y2[i], level,0,sign);
    end;
    ch := ReadKey;
end.
```

8

The Hilbert Curve

The Hilbert curve is one of the Peano family of curves, but it has some subtle differences that make it unique. Figure 8-1 shows the generator and the next level of the Hilbert curve. Since we're used to pretty straightforward applications of the generator to the line segments of the initiator or previous level of the curve, it may be quite difficult to visualize what is happening with the Hilbert curve. The parameters that we use are:

$$r = 1/2 \qquad \text{(Equation 8-1)}$$

and

$$N = 4 \qquad \text{(Equation 8-2)}$$

In other words, the line segment of the generator is one-half of the line segment to which it is being applied, and the generator is applied four times. But to make things more complex, each time that we go to a lower level to run the generator program, we return to the next higher level and draw a line with the same length being used at the lower level. This sounds a little complex. Try tracing it out on Figure 8-1(b). Of course, each time you run the generator program, you have to make sure that it has the proper orientation for the curve to come out correctly. Starting at the lower right of Figure 8-1(a) we use the generator, then draw a line segment to the left, then use the generator again, then draw a line segment up, then use the generator again,

then draw a line segment to the right, and then use the generator one
final time.

Figure 8-1. Generator and Second Level for Hilbert Curve

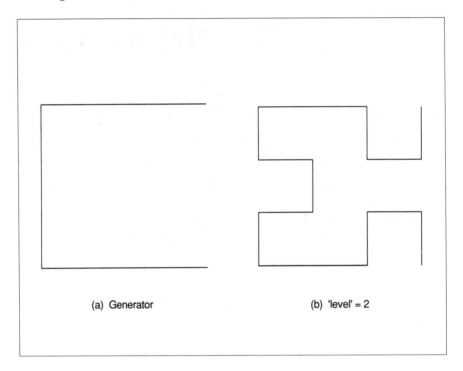

(a) Generator   (b) 'level' = 2

Generating the Hilbert Curve

All of the above becomes quite clear from the program listing of
Figure 8-2. This program is quite similar to the generic initiator/gen-
erator program of Chapter 6, but since all of our steps are either in
the horizontal or vertical directions, we don't use the turtle graphics
to keep track of direction, but simply step in the proper direction at
each operation. You will note that at each level of *generate* except
the lowest, we both call *generate* recursively a number of times and
also draw line segments of the proper length between calls to

generate. The resulting curves for levels from 3 to 6 are shown in Figure 8-3.

It is worth noting that there is an entirely different way to generate the Hilbert curve. The program to do this is listed in Figure 8-4. It makes use of four separate functions to do the generate tasks, and, although elegant, tends to obscure what is going on.

Figure 8-2. Program to Generate Hilbert Curve

```
program hilbert;

uses CRT,Graph,Fractal;

var
    graphDriver,GraphMode,i,generator_size,level,init_size:
        integer;
    x1,x2,y1,y2,r,r1,r2,temp: real;
    ch: char;


procedure generate (r1: real; r2: real);

    begin
        dec(level);
        if level > 0 then
            generate(r2,r1);
        x2 := x2 + r1;
        y2 := y2 + r2;
        Line(Round(x1+320),Round(175 - y1*0.729),
            Round(x2+320),Round(175 - y2*0.729));
        x1 := x2;
        y1 := y2;
        if level > 0 then
            generate(r1,r2);
        x2 := x2 + r2;
        y2 := y2 + r1;
        Line(Round(x1+320),Round(175 - y1*0.729),
            Round(x2+320),Round(175 - y2*0.729));
        x1 := x2;
        y1 := y2;
        if level > 0 then
            generate(r1,r2);
```

```
      x2 := x2 - r1;
      y2 := y2 - r2;
      Line(Round(x1+320),Round(175 - y1*0.729),
          Round(x2+320),Round(175 - y2*0.729));
      x1 := x2;
      y1 := y2;              if level > 0 then
          generate(-r2,-r1);
      inc(level);
   end;

begin

   write('Enter level (1 - 8): ');
   readln(level);
   if level < 1 then
       level := 1;
   GraphDriver := 4;
   GraphMode := EGAHi;
   InitGraph(graphDriver,GraphMode,'');
   SetLineStyle(0,$FFFF,1);
   SetColor(15);
   temp := 2;
   i := level;
   while i>1 do
       begin
           temp := temp*2;
           dec(i);
       end;
   r := 400/temp;
   x1 := -200;
   y1 := -200;
   x2 := -200;
   y2 := -200;
   generate(r,0);
   ch := ReadKey;
end.
```

Figure 8-3. Hilbert Curve for Levels 3 to 6

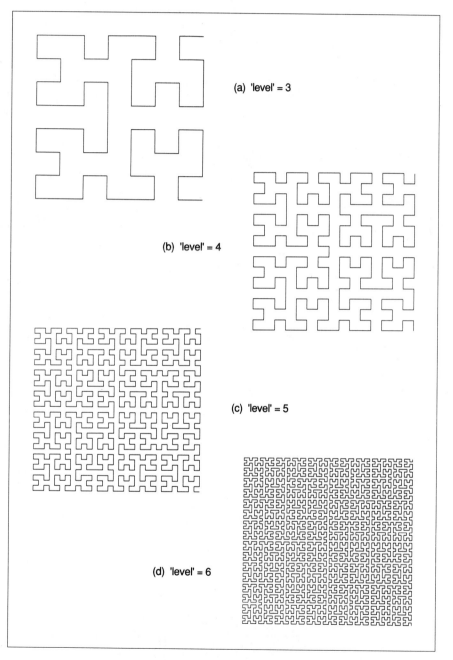

(a) 'level' = 3

(b) 'level' = 4

(c) 'level' = 5

(d) 'level' = 6

Figure 8-4. Alternate Program to Draw Hilbert Curve

```pascal
program hilbert2;

uses CRT,Graph,Fractal;

var
     graphDriver,GraphMode,i,j,level: integer;
     x,y,old_x,old_y,h: integer;
     ch: char;

procedure gen2(i: integer); forward;
procedure gen3(i: integer); forward;
procedure gen4(i: integer); forward;

procedure gen1(i: integer);
    begin
        if i > 0 then
        begin
            gen4(i-1);
            x := x - h;
             Line(old_x+320,175-(old_y*35) div
                 48,x+320,175-(y*35) div 48);
            old_x := x;
            old_y := y;
            gen1(i-1);
            y := y - h;
             Line(old_x+320,175-(old_y*35) div
                 48,x+320,175-(y*35) div 48);
            old_x := x;
            old_y := y;
            gen1(i-1);
            x := x + h;
             Line(old_x+320,175-(old_y*35) div
                 48,x+320,175-(y*35) div 48);
            old_x := x;
            old_y := y;
            gen2(i-1);
        end;
    end;
  procedure gen2(i: integer);
    begin
        if i > 0 then
        begin
            gen3(i-1);
            y := y + h;
             Line(old_x+320,175-(old_y*35) div
                 48,x+320,175-(y*35) div 48);
            old_x := x;
            old_y := y;
            gen2(i-1);
            x := x + h;
             Line(old_x+320,175-(old_y*35) div
                 48,x+320,175-(y*35) div 48);
            old_x := x;
```

```
                old_y := y;
                gen2(i-1);
                y := y - h;
                 Line(old_x+320,175-(old_y*35) div
                     48,x+320,175-(y*35) div 48);
                old_x := x;
                old_y := y;
                gen1(i-1);
            end;
        end;

procedure gen3(i: integer);
    begin
        if i > 0 then
        begin
            gen2(i-1);
            x := x + h;
             Line(old_x+320,175-(old_y*35) div
                 48,x+320,175-(y*35) div 48);
            old_x := x;
            old_y := y;
            gen3(i-1);
            y := y + h;
             Line(old_x+320,175-(old_y*35) div
                 48,x+320,175-(y*35) div 48);
            old_x := x;
            old_y := y;
            gen3(i-1);
            x := x - h;
             Line(old_x+320,175-(old_y*35) div
                 48,x+320,175-(y*35) div 48);
            old_x := x;
            old_y := y;
            gen4(i-1);
        end;
    end;

procedure gen4(i: integer);
    begin
        if i > 0 then
        begin
            gen1(i-1);
            y := y - h;
             Line(old_x+320,175-(old_y*35) div
                 48,x+320,175-(y*35) div 48);
            old_x := x;
            old_y := y;
            gen4(i-1);
            x := x - h;
             Line(old_x+320,175-(old_y*35) div
                 48,x+320,175-(y*35) div 48);
            old_x := x;
            old_y := y;
            gen4(i-1);
            y := y + h;
             Line(old_x+320,175-(old_y*35) div
```

```
                48,x+320,175-(y*35) div 48);
            old_x := x;
            old_y := y;
            gen3(i-1);
        end;
    end;

begin
    write('Enter level (1 - 7): ');
    readln(level);
    if level < 1 then
        level := 1;
    GraphDriver := 4;
    GraphMode := EGAHi;
     InitGraph(graphDriver,GraphMode,'');
     SetLineStyle(0,$FFFF,1);
     SetColor(15);
     h := 448;
     x := 0;
     y := 0;
     for i:=1 to level do
     begin
         h := h div 2;
         x := x + h div 2;
         y := y + h div 2;
         old_x := x;
         old_y := y;
     end;
     gen1(level);
     ch := ReadKey;
end.
```

Fractal Dimension of the Hilbert Curve

If you take a close look at the Hilbert curve, it is evident that after a sufficiently large number of iterations, it will pass through every point in the plane. Going back to the formula for fractal dimension, we have:

$$D = \log(4) \ / \ \log(2) = 2 \qquad\qquad \text{(Equation 8-2)}$$

This confirms that the Hilbert curve is a Peano Curve, and that it passes through every point on the plane.

Hilbert Curve in Three Dimensions

The Hilbert curve can also be drawn in higher dimensions, but it becomes rather difficult to determine the proper orientations of the generator to assure that every point is covered without duplication. Figure 8-5 shows the second level of a three-dimensional curve. The program to draw these two displays is listed in Figure 8-5. This program breaks down for higher levels, since we haven't found the proper orientations to insert to assure that they would be correct. You're welcome to hunt for these if you want to, but they aren't obvious.

Figure 8-5. Three-Dimensional Hilbert Curve

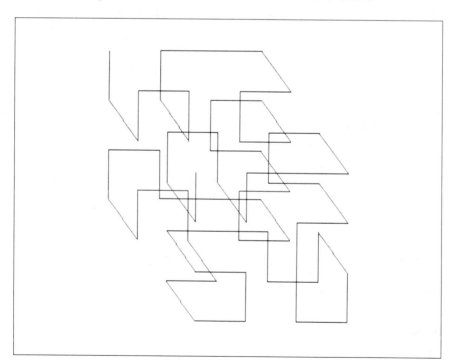

Using the Hilbert Curve for Display Data Storage

In Chapter 3, we looked at a program for compressing display information and storing it in a disk file for recovery and redisplay at a later time. The run length encoding method substantially compressed the data file by requiring only two bytes to define up to 63 pixels in a color plane if they are all alike. This is not the optimum run length recording for two reasons. The first is that we are limited to 63 pixels because we only allowed six bits to define the number of pixels affected.

This was done to allow the two most significant bits to be a flag which indicates that the byte represents a number rather than a data byte. This whole scheme, in this case, is constrained by the fact that we are trying to work with bytes; in a word-oriented system, we would have more bits available to work with. The second reason that we don't have optimum compression is that we record one line at a time from each of the three color planes. Even if there is a large block of the same color, it is unlikely that the pixel data from one color plane line will be the same as that from the next.

Now suppose that the color data is represented by a three-dimensional volume in which the x and y dimensions are the same as they were for the display, but the third dimension represents color. We would like to record this as a long string in a single file. One way to do this is to scan through each plane, line by line. But, as pointed out above, when we move from one line to the next, any continuity of color data that might let us use maximum run length compression is lost. What we need is a way of scanning through the three-dimensional space which will give us a one-dimensional result in which points that were close together in the original space will still be close together on the resulting line. Thus, a block of a single color on the original display will be lumped together on the resulting line, and is suitable for compression to a few bytes.

The Hilbert curve performs exactly this function. It scans an n-dimensional surface and reduces it to a one-dimensional line, and it has the characteristic that points that are close together on the n-dimensional surface are close together on the resulting line. Of course there is some loss of information on the closeness of points because a single dimension cannot possibly have the same degree of spatial associativity that can be achieved with a higher dimension of space. However, this loss is minimal compared with other techniques that might be used for transforming the data. F. H. Preston, A. F. Lehar, and R. J. Stevens of the S. R. D. B. Home Office in England have developed algorithms for using the Hilbert curve to map image data and for compressing the resulting information. They insist on calling the Hilbert curve a "Peano Curve," which is unfortunate, since, as we have already discovered, there is a whole family of Peano curves, of which the Hilbert curve is only a single specific type. They have published several papers on their results, one of which is referenced in the Bibliography.

Figure 8-6. Program to Draw Three-Dimensional Hilbert Curve

```
program hil3d;

uses CRT,Graph,Fractal;

var
    GraphDriver,GraphMode,i,level: integer;
    points: array[0..2] of real;
    temp,x1,x2,y1,y2,r,x_angle,y_angle,z_angle,cx,cy,cz,
        sx,sy,sz: real;
    ch: char;


procedure generate (a: integer; b: integer; c: integer);

    var
        sign: array[0..2] of integer;
    begin
        sign[0] := 1;
        sign[1] := 1;
        sign[2] := 1;
        dec(level);
        if a < 0 then
            sign[0] := -1;
```

```
a := Abs(a)-1;
if b < 0 then
    sign[1] := -1;
b := Abs(b)-1;
if c < 0 then
    sign[2] := -1;
c := Abs(c)-1;
x1 := points[0]*cx + points[1]*cy + points[2]*cz;
y1 := points[0]*sx + points[1]*sy + points[2]*sz;
if level > 0 then
    generate(-2,1,3);
points[a] := points[a] + (r*sign[0]);
x2 := points[0]*cx + points[1]*cy + points[2]*cz;
y2 := points[0]*sx + points[1]*sy + points[2]*sz;
Line(Round(x1+320),Round(175 - y1*0.729)
    ,Round(x2+320),Round(175 - y2*0.729));
x1 := points[0]*cx + points[1]*cy +points[2]*cz;
y1 := points[0]*sx + points[1]*sy + points[2]*sz;
 if level > 0 then
    generate(3,1,-2);
points[b] := points[b] + (r*sign[1]);
x2 := points[0]*cx + points[1]*cy + points[2]*cz;
y2 := points[0]*sx + points[1]*sy + points[2]*sz;
Line(Round(x1+320),Round(175 - y1*0.729)
    ,Round(x2+320),Round(175 - y2*0.729));
x1 := points[0]*cx + points[1]*cy + points[2]*cz;
y1 := points[0]*sx + points[1]*sy + points[2]*sz;
if level > 0 then
    generate(3,1,-2);
points[a] := points[a] - (r*sign[0]);
x2 := points[0]*cx + points[1]*cy + points[2]*cz;
y2 := points[0]*sx + points[1]*sy + points[2]*sz;
 Line(Round(x1+320),Round(175 - y1*0.729)
    ,Round(x2+320),Round(175 - y2*0.729));
x1 := points[0]*cx + points[1]*cy + points[2]*cz;
y1 := points[0]*sx + points[1]*sy + points[2]*sz;
if level > 0 then
    generate(2,-3,1);
points[c] := points[c] + (r*sign[2]);
x2 := points[0]*cx + points[1]*cy + points[2]*cz;
y2 := points[0]*sx + points[1]*sy + points[2]*sz;
Line(Round(x1+320),Round(175 - y1*0.729)
    ,Round(x2+320),Round(175 - y2*0.729));
x1 := points[0]*cx + points[1]*cy + points[2]*cz;
y1 := points[0]*sx + points[1]*sy + points[2]*sz;
 if level > 0 then
    generate(-3,1,2);
points[a] := points[a] + (r*sign[0]);
x2 := points[0]*cx + points[1]*cy + points[2]*cz;
y2 := points[0]*sx + points[1]*sy + points[2]*sz;
 Line(Round(x1+320),Round(175 - y1*0.729)
```

```
          ,Round(x2+320),Round(175 - y2*0.729));
      x1 := points[0]*cx + points[1]*cy + points[2]*cz;
      y1 := points[0]*sx + points[1]*sy + points[2]*sz;
       if level > 0 then
           generate(-2,3,1);
      points[b] := points[b] - (r*sign[1]);
      x2 := points[0]*cx + points[1]*cy + points[2]*cz;
      y2 := points[0]*sx + points[1]*sy + points[2]*sz;
       Line(Round(x1+320),Round(175 - y1*0.729)
           ,Round(x2+320),Round(175 - y2*0.729));
      x1 := points[0]*cx + points[1]*cy + points[2]*cz;
      y1 := points[0]*sx + points[1]*sy + points[2]*sz;
       if level > 0 then
           generate(3,-1,2);
      points[a] := points[a] - (r*sign[0]);
      x2 := points[0]*cx + points[1]*cy + points[2]*cz;
      y2 := points[0]*sx + points[1]*sy + points[2]*sz;
       Line(Round(x1+320),Round(175 - y1*0.729)
           ,Round(x2+320),Round(175 - y2*0.729));
      x1 := points[0]*cx + points[1]*cy + points[2]*cz;
      y1 := points[0]*sx + points[1]*sy + points[2]*sz;
      if level > 0 then
      generate(-2,-1,-3);
      inc(level);
   end;
begin
   x_angle := -55;
   y_angle := 90;
   z_angle := 0;
   write('Enter level (1 - 8): ');
   readln(level);
   if level < 1 then
       level := 1;
   GraphDriver := 4;
   GraphMode := EGAHi;
   InitGraph(graphDriver,GraphMode,'');
   SetLineStyle(0,$FFFF,1);
   SetColor(15);
   sx := Sin(x_angle*0.017453292);
   sy := Sin(y_angle*0.017453292);
   sz := Sin(z_angle*0.017453292);
   cx := Cos(x_angle*0.017453292);
   cy := Cos(y_angle*0.017453292);
   cz := Cos(z_angle*0.017453292);
   temp := 2;
   i := level;
   while i>1 do
       begin
           temp := temp*2;
           dec(i);
       end;
```

```
    r := 300/temp;
    points[0] := -200;
    points[1] := 50;
    points[2] := 0;
    generate(3,-2,1);
    ch := ReadKey;
end.
```

9

The Sierpinski Curve

The Sierpinski curve is particularly interesting because there are several ways of generating it that seem to start with quite different premises but end up producing essentially the same curve, and also because it has practical uses for space-filling required by clustering algorithms used in route optimization. The first method of generating the Sierpinski triangle is that which we are most familiar with, namely the use of the initiator/generator technique first described in Chapter 6. For this curve, the initiator is a straight line. The generator for the curve and the resulting curve for levels 2 and 3 are shown in Figure 9-1. Curves for levels 4, 6, and 8 are shown in Figure 9-2. It's not a very good idea to carry the curve to higher levels than 8, since the triangles begin to fill in too much and detail is lost. The program to generate the Sierpinski triangle is listed in Figure 9-3.

Figure 9-1. Sierpinski Triangles

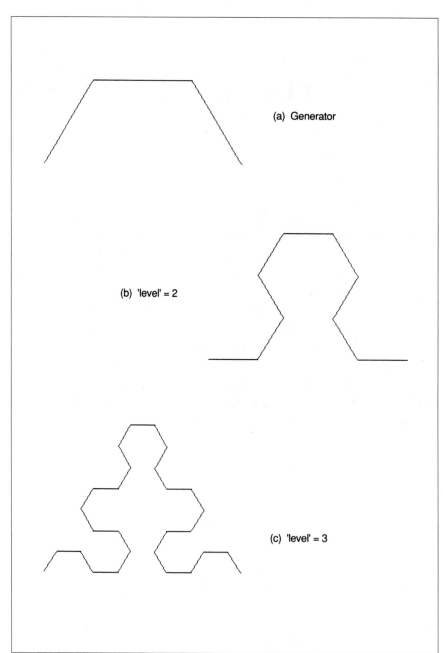

(a) Generator

(b) 'level' = 2

(c) 'level' = 3

Figure 9-2. Higher Levels of Sierpinski Triangles

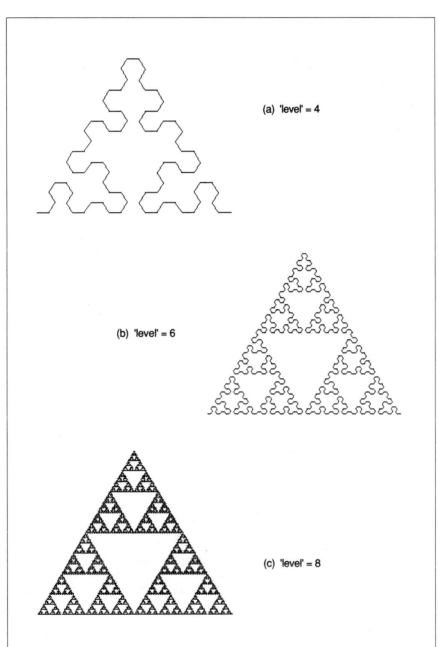

(a) 'level' = 4

(b) 'level' = 6

(c) 'level' = 8

Figure 9-3. Program to Generate Sierpinski Triangles

```
program sierp;

uses CRT,Graph,Fractal;

var
    graphDriver,GraphMode,i,generator_size,level: integer;
    Xpoints, Ypoints: array[0..24] of real;
    ch: char;


procedure generate (X1: real; Y1: real; X2: real; Y2: real;
    level: integer; sign: integer);
    var
        j,k,int_sign: integer;
        Xpoints, Ypoints: array[0..24] of real;

    begin
        turtle_r := sqrt((X2 - X1)*(X2 - X1) + (Y2 - Y1)*(Y2
            - Y1))/2.0;
        turtle_x := X1;
        turtle_y := Y1;
        Xpoints[0] := X1;
        Ypoints[0] := Y1;
        Xpoints[3] := X2;
        Ypoints[3] := Y2;
        turtle_theta := point(X1,Y1,X2,Y2);
        turn(60*sign);
        step;
        Xpoints[1] := turtle_x;
        Ypoints[1] := turtle_y;
        turn(-60*sign);
        step;
        Xpoints[2] := turtle_x;
        Ypoints[2] := turtle_y;
        dec(level);
        sign := -sign;
        if level = 0 then
        begin
            for k:=0 to generator_size - 1 do
            begin
                Line(Round(Xpoints[k]+320),Round(175 -
                    Ypoints[k]*0.729)
                    ,Round(Xpoints[k+1]+320),Round(175
                    - Ypoints[k+1]* 0.729));
            end;
        end
        else
        begin
```

```
            int_sign := sign;
            for j:=0 to generator_size-1 do
            begin
                X1 := Xpoints[j];
                X2 := Xpoints[j+1];
                Y1 := Ypoints[j];
                Y2 := Ypoints[j+1];
                generate (X1,Y1,X2,Y2,level,int_sign);
                int_sign := -int_sign;
            end;
        end;
    end;

begin
    generator_size := 3;
    write('Enter level (1 - 8): ');
    readln(level);
    if level < 1 then
        level := 1;
    GraphDriver := 4;
    GraphMode := EGAHi;
    InitGraph(graphDriver,GraphMode,'');
    SetLineStyle(0,$FFFF,1);
    SetColor(15);
    generate(-130,0,130,0, level, 1);
    ch := ReadKey;
end.
```

Sierpinski Gasket

As one looks at the triangle of Figure 9-2(c), one gets the idea that this curve could be produced by starting with one big filled-in triangle and cutting out smaller and smaller ones from it in appropriate places. This is the technique that is used in the program listed in Figure 9-4.

We first create and fill a large triangle, using the Turbo Pascal procedure *FillPoly.* You will note that Turbo Pascal has a peculiar structure for defining the points that make up the triangle (or any polygon to be filled). Interestingly enough, many Turbo Pascal books tell you how to initiate these peculiar values but don't explain how to change them during the program. If you look at the listing of Figure 9-4, the technique is obvious.

Once the triangle is drawn and filled, the program calls the function *generate*, which divides the triangle into four smaller ones. The *generate* function then calls the function *node*. This function blanks out the center triangle (by using the function *FillPoly* with the color black) and then calls *generate* (in a recursion process) to operate upon the three peripheral triangles. The procedure continues to whatever level you have entered into the parameter level.

Note that the technique of removing triangles has the drawback that if you use too high a level, there is insufficient display resolution to preserve the colored portions of the display, with the result that the entire original triangle is eventually blanked out. The program to perform this operation is listed in Figure 9-4. The resulting Sierpinski triangle is shown in Figure 9-5.

Figure 9-4. Program to Generate Sierpinksy Gasket

```
program siergask;

uses CRT,Graph,Fractal;

var
    Triangle: array[1..3] of PointType;
    graphDriver,GraphMode,i,level,L_length,x1,y1,x2,y2,x3,y3:
        integer;
    ch: char;

procedure node(x1: integer; y1: integer; x2: integer; y2:
    integer; x3: integer; y3: integer; x4: integer; y4:
    integer; x5: integer; y5: integer; x6: integer; y6:
    integer; level: integer; l_length: integer); forward;

procedure generate(x1: integer; y1: integer; x2: integer;
    y2: integer; x3: integer; y3: integer; level: integer;
    l_length: integer);

    var
        line_length,x4,y4,x5,y5,x6,y6: integer;

    begin
        line_length := l_length div 2;
        x4 := x1 + line_length;
        y4 := y1;
        x5 := x1 + line_length div 2;
        y5 := y1 - Round((line_length/2)*1.26292);
        x6 := x5 + line_length;
        y6 := y5;
```

```
                node(x1,y1,x2,y2,x3,y3,x4,y4,x5,y5,x6,y6,
                    level,line_length);
        end;

    procedure node(x1: integer; y1: integer; x2: integer; y2:
        integer; x3: integer; y3: integer; x4: integer; y4:
        integer; x5: integer; y5: integer; x6: integer; y6:
        integer; level: integer; l_length: integer);
        begin
            Triangle[1].x := x4;
            Triangle[1].y := y4;
            Triangle[2].x := x5;
            Triangle[2].y := y5;
            Triangle[3].x := x6;
            Triangle[3].y := y6;
            FillPoly(3,Triangle);
            if level <> 0 then
            begin
                generate (x1,y1,x4,y4,x5,y5,level-1,l_length);
                generate (x4,y4,x2,y2,x6,y6,level-1,l_length);
                generate (x5,y5,x6,y6,x3,y3,level-1,l_length);
            end;
        end;

    begin
        write('Enter level (1 - 5): ');
        readln(level);
        if level < 1 then
            level := 1;
        GraphDriver := 4;
        GraphMode := EGAHi;
        InitGraph(graphDriver,GraphMode,'');
        x1 := 64;
        y1 := 335;
        x2 := 576;
        y2 := 335;
        x3 := 320;
        y3 := 15;
        l_length := 512;
        SetFillStyle(1,15);
        Triangle[1].x := x1;
        Triangle[1].y := y1;
        Triangle[2].x := x2;
        Triangle[2].y := y2;
        Triangle[3].x := x3;
        Triangle[3].y := y3;
        FillPoly(3,Triangle);
        SetFillStyle(1,0);
        SetColor(0);
        generate(x1,y1,x2,y2,x3,y3,level,l_length);
        ch := ReadKey;
    end.
```

Figure 9-5. Sierpinski Gasket

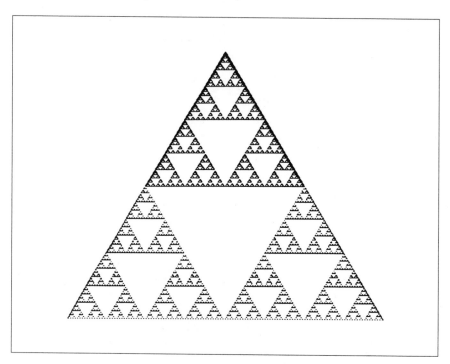

Other Methods of Generating
the Sierpinski Triangle

There is another method of generating the Sierpinski triangle, which makes use of an algorithm similar to that used for generating strange attractors in Chapter 4, and for IFS systems in Chapter 20. The program listing is given in Figure 9-6. It starts out with a point at a random location on the screen. It then randomly selects one of three transformations. The first simply creates a new point at half the x and y coordinates of the previous point. The second creates a new point whose x coordinate is the previous x plus 639 (distance across the display) and divided by two, and whose y coordinate is half the previous y coordinate value. The third transformation creates a new point whose x coordinate is the previous x plus 320 (half the distance across the display) and divided by two and whose y coordinate is the previous y coordinate plus 349 (distance down the display) and divided by two. The result of plotting 120,000 points is the Sierpinski triangle, as shown in Figure 9-7(a). Another, similar method of generating the Sierpinski triangle is through the use of iterated function systems as explained in Chapter 20. A very short and simple code describes the triangle to the IFS and results in a good representation being drawn. For further details, refer to Chapter 20.

Figure 9-6. Another Program to Generate the Sierpinski Triangle

```
program sirchet3;

uses CRT,Graph;

var
    graphDriver,GraphMode,x,y,switcher: integer;
    i: longint;
    ch: char;

begin
    GraphDriver := 4;
    GraphMode := EGAHi;
    InitGraph(graphDriver,GraphMode,'');
    x := random(640);
    y := random(350);
    for i:=0 to 120000 do
    begin
        switcher := random(3);
        case switcher of

            0:      begin
                        x := x div 2;
                        y := y div 2;
                    end;
            1:      begin
                        x := (x + 639) div 2;
                        y := y div 2;
                    end;
            2:      begin
                        x := (x + 320) div 2;
                        y := (y + 349) div 2;
                    end;
        end;
        if i > 20 then
            PutPixel(x,y,15);
    end;
    ch := ReadKey;
end.
```

Figure 9-7. Sierpinski Triangle and Cousins

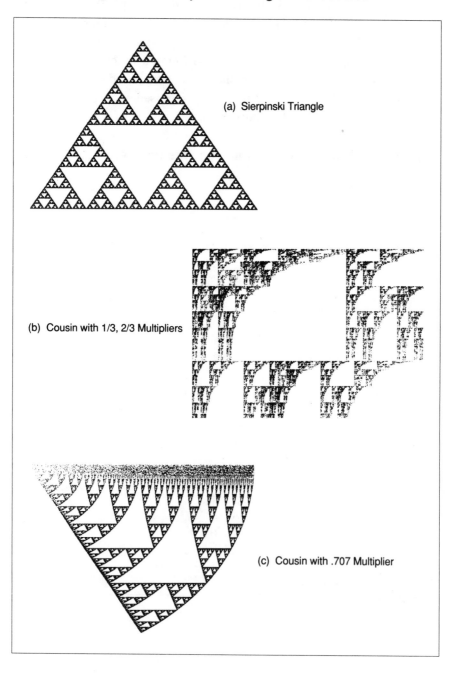

(a) Sierpinski Triangle

(b) Cousin with 1/3, 2/3 Multipliers

(c) Cousin with .707 Multiplier

Strange Cousins of the Sierpinski Triangle

I am indebted to my friend Chester Stromswold for pointing out to me the strange cousins of the Sierpinski triangle which can be generated by slight modifications of the above program. The first variation uses multipliers of 2/3 or 1/3 instead of 1/2 throughout the program. The program to generate this figure is shown in Figure 9-8. The resulting figure is shown in Figure 9-7(b). The second variation uses a multiplier of 0.7071068 (reciprocal of the square root of two) instead of 1/2 at several critical places. The program to generate this figure is listed in Figure 9-9. The resulting figure is shown in Figure 9-7(c).

Figure 9-8. Program to Generate Cousin of the Sierpinski Triangle with 1/3 and 2/3 Multipliers

```
program sierchet;

uses CRT,Graph;

var
    graphDriver,GraphMode,x,y,switcher: integer;
    i: longint;
    ch: char;

begin
    GraphDriver := 4;
    GraphMode := EGAHi;
    InitGraph(graphDriver,GraphMode,'');
    x := random(640);
    y := random(350);
    for i:=0 to 120000 do
    begin
        switcher := random(4);
        case switcher of

            0:    begin
                    x := x div 3;
                    y := y div 3;
                  end;
            1:    begin
                    x := (x + 639)*2 div 3;
                    y := y div 3;
                  end;
            2:    begin
```

```
                    x := (x + 639) div 3;
                    y := (y + 349)*2 div 3;
                 end;
           3:    begin
                    x := x div 3;
                    y := (y + 349)*2 div 3;
                 end;
        end;          if i > 20 then
             PutPixel(x,y,15);
    end;
    ch := ReadKey;
end.
```

Figure 9-9. Program to Generate Cousin of the Sierpinski Triangle Using .7071068 Multipliers

```
program sirchet2;

uses CRT,Graph;

var
    graphDriver,GraphMode,switcher: integer;
    i,x,y: longint;
    ch: char;
    s2: real;

begin
    GraphDriver := 4;
    GraphMode := EGAHi;
    InitGraph(graphDriver,GraphMode,'');
    s2 := sqrt(0.5);
    x := random(640);
    y := random(350);
    for i:=0 to 120000 do
    begin
        switcher := random(3);
        case switcher of

            0:    begin
                    x := Round(x*s2);
                    y := Round(y*s2);
                  end;
            1:    begin
                    x := Round(sqrt((639*639 + x*x)/2));
                    y := Round(y*s2);
                  end;          2:    begin
                    x := Round(sqrt((320*320 + x*x)/2));
                    y := Round(sqrt((349*349 + y*y)/2));
                  end;
```

```
        end;
        if i > 20 then
            PutPixel(x,y,15);
    end;
    ch := ReadKey;
end.
```

Sierpinski Box

The same technique described above for the Sierpinski gasket can be applied to create a rectangular figure, which I have called a Sierpinski box. The program to create the box is listed in Figure 9-10 and the resulting figure is shown in Figure 9-11. The program is much like that described previously. The Turbo Pascal procedure *FillPoly* is again used, but this time with a structure that includes four coordinate points, to define a square. Once a large square is created and filled, it is divided into nine smaller squares by a new version of the function *generator*. A new version of the function node is then used to blank out the center square and then call generate for each of the eight peripheral squares. Again, caution must be used in selecting the value of the parameter level. If level is too large, lack of display resolution will cause the entire original square to be blanked out.

Figure 9-10. Program to Generate Sierpinski Box

```
program sierbox;

uses CRT,Graph,Fractal;

var
    Square: array[1..4] of PointType;
    graphDriver,GraphMode,i,level,L_length,x1,y1,x2,y2,x3,y3:
        integer;
    ch: char;

procedure node(x1: integer; y1: integer; x2: integer; y2:
    integer; x3: integer; y3: integer; x4: integer; y4:
    integer; level: integer; l_length: integer); forward;

procedure generate(x1: integer; y1: integer; x2: integer; y2:
    integer; level: integer; l_length: integer);
```

```pascal
var
    line_length,x3,y3,x4,y4: integer;

    begin
        line_length := l_length div 3;
        x3 := x1 + line_length;
        y3 := y1 - (35*line_length) div 48;
        x4 := x2 - line_length;
        y4 := y2 + (35*line_length) div 48;
        node (x1,y1,x2,y2,x3,y3,x4,y4,level,line_length);
    end;

procedure node(x1: integer; y1: integer; x2: integer; y2:
    integer; x3: integer; y3: integer; x4: integer; y4:
    integer; level: integer; l_length: integer);

    begin
        Square[1].x := x3;
        Square[1].y := y3;
        Square[2].x := x4;
        Square[2].y := y3;
        Square[3].x := x4;
        Square[3].y := y4;
        Square[4].x := x3;
        Square[4].y := y4;
        FillPoly(4,Square);
        if level <> 0 then
        begin
            generate (x1,y1,x3,y3,level-1,l_length);
            generate (x3,y1,x4,y3,level-1,l_length);
            generate (x4,y1,x2,y3,level-1,l_length);
            generate (x1,y3,x3,y4,level-1,l_length);
            generate (x4,y3,x2,y4,level-1,l_length);
            generate (x1,y4,x3,y2,level-1,l_length);
            generate (x3,y4,x4,y2,level-1,l_length);
            generate (x4,y4,x2,y2,level-1,l_length);
        end;
    end;

begin
    write('Enter level (1 - 3): ');
    readln(level);
    if level < 1 then
        level := 1;
    GraphDriver := 4;
    GraphMode := EGAHi;
    InitGraph(graphDriver,GraphMode,'');
    x1 := 100;
    y1 := 335;
    x2 := 540;
```

```
        y2 := 15;
        l_length := 440;
        SetFillStyle(1,15);
        Square[1].x := x1;
        Square[1].y := y1;
        Square[2].x := x2;
        Square[2].y := y1;
        Square[3].x := x2;
        Square[3].y := y2;
        Square[4].x := x1;
        Square[4].y := y2;
        FillPoly(4,Square);
        SetFillStyle(1,0);
        SetColor(0);
        generate(x1,y1,x2,y2,level,l_length);
        ch := ReadKey;
end.
```

Figure 9-11. Sierpinski Box

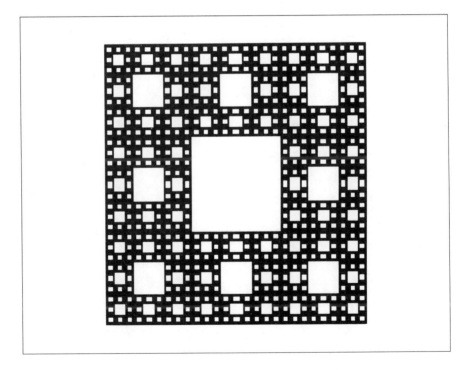

10

Trees

In the past few chapters, we have created fractal curves by repeatedly replacing line segments with scaled-down replicas of a generator pattern. The result has been curves that were self-similar, a blown-up version of a small section of the curve has a very similar shape to that of a larger portion of the curve. Now, we are going to take a different approach. We will start with a stem; at its end we will to branch off in two directions and draw two branches. We will repeatedly perform this process at the end of each new branch. The result is a tree. Since one of the purposes of this exercise is to use these curves to represent trees in nature, we first need to discuss something about real trees.

Real Trees

The rough outline of the tree creation process given above implied that at each node in the tree creating process, we would branch off in two directions. The result will be a two-dimensional tree, but hopefully it will have some relation to real three-dimensional trees. Before going further, step outside and look at a few real trees. The first thing to note is that there are two classes of trees, deciduous (trees whose leaves fall every year) and conifers (evergreens having cones), and that these two classes of trees are built quite differently. The conifers tend to have rings of branches at different heights around a central trunk. This does not seem to square at all with the binary branching process, and indeed, we will see later that the tree curves that we generate never look like conifers.

The second thing to note is that the deciduous trees, although they are closer in appearance to our model, still are much more complex in their structure. While binary branching is often the rule, there are exceptions, where a stem splits into more than two branches. Furthermore, the lengths of stems before branching occurs differ randomly from the norm, as do the diameters of branches. The reason for making a point of all this is that we are next going to present some data on expressions for modeling trees, but we want to make sure that these are not taken as true representations of the way real trees are constructed. In some literature, authors appear to have been overpowered by their ability to express tree structures mathematically, to the point that the model supersedes reality. Remember, the mathematical formulas are a nice way of generating tree curves, but the real tree is much more complex, and much more interesting. If you want a real challenge, take the tree program that we will list later, and attempt to expand it to cover each of the possible situations for a real tree.

Mathematical Representation of Trees

Everyone seems to be fond of quoting Leonardo da Vinci's observation to the effect that the sum of the cross-sectional areas of all tree branches at a given height is constant. This should not be too surprising; the tree is required to pass nutrients from the roots to the leaves and for a given nutrient requirement one might expect that the "pipe" cross-sectional area required for nutrient transportation would be constant, regardless of height or the number of "pipes." When we translate this observation to diameters (or widths when we make our two-dimensional drawings) we have an expression of the form:

$$D_0^\alpha = D_1^\alpha + D_2^\alpha \qquad \text{(Equation 10-1)}$$

where D_0 is the diameter of the stem, D_1 and D_2 are the diameters of the two branches that the stem splits into, and α is 2 according to da Vinci. There are other forms of tree-like structures. The simple model given above probably applies better to river networks than to trees, since the likelihood that more than two tributaries of a river system would join at the same place is remote. Other trees are found in the human body in the form of the arterial blood transportation system and the bronchi. Investigations have shown that a good approximation for α for the bronchial system is 3, and for the arteries is 2.7.

When we come to construct our program for tree generation, we shall use the expression:

$$B_{n+1} = 2^{-1/\alpha} B_n \qquad \text{(Equation 10-2)}$$

where B_n is the diameter of the lower-level branch and B_{n+1} represents the diameter of each of the two branches into which B_n splits. We also need to consider the length of the branches. McMahon studied various typical trees and concluded that a similar recursive formula for length could be written as:

$$L_{n+1} = 2^{-3/(2\alpha)} L_n \qquad \text{(Equation 10-3)}$$

where L_n is the length of the predecessor branch and L_n+1 is the length of each of the two successor branches after bifurcation.

Tree-Drawing Program

Figure 10-1 lists a program for drawing trees. It permits entering the initial length and width of the stem, the value of α for the left and right sides of the tree, the left and right branching angles, and the level of recursion. You will note that the program is a lot like those we have been using in the previous few chapters. It first computes the right and left width and length factors using equations 10-2 and 10-3.

Next it sets up the parameters for the beginning and end of the stem and its width and draws it. The *turtle_theta* parameter is then set up to point in the direction of the stem, and then turned to the left angle. The function *generate* is then run recursively until the lowest level is reached, then the *turtle_theta* parameter is reset in the stem direction and turned through the right angle and the generate function run again. Note that the height and width parameters passed to the generate function are scaled down by multiplying by the appropriate scale factors at the time of function call.

Figure 10-1. Program to Generate Trees

```
program trees;

uses CRT, Graph,Fractal;

const
    ln_2: real = (0.6931471);

var
    Square: array[1..4] of PointType;
    GraphDriver,GraphMode,i,j,level,width: integer;
    height,left_alpha,right_alpha,left_angle,
        right_angle, left_width_factor,left_height_factor,
        right_width_factor,right_height_factor,x,y,x1,
        y1,x2,y2: real;
    ch: char;

procedure generate(x: real; y: real; width: integer; height:
    real; angle: real; level: integer);

    var
        x1,y1: real;

    begin
        turtle_x := x;
        turtle_y := y;
        turtle_r := height;
        step;
        x1 := turtle_x;
        y1 := turtle_y;
        dec(level);
        if level < 3 then
        begin
            SetColor(10);
            SetFillStyle(1,10);
```

```
            end
            else
            begin
                SetColor(6);
                SetFillStyle(1,6);
            end;
            if Abs(x - x1) > Abs(y - y1) then
            begin
                Square[1].x := Round(x+320);
                Square[1].y := Round(175-y*0.729) + width div 2;
                Square[2].x := Round(x+320);
                Square[2].y := Round(175-y*0.729) - width div 2;
                Square[3].x := Round(x1+320);
                Square[3].y := Round(175-y1*0.729)-width div 2;
                Square[4].x := Round(x1+320);
                Square[4].y := Round(175-y1*0.729)+width div 2;
            end
            else
            begin
                Square[1].x := Round(x+320) - width div 2;
                Square[1].y := Round(175 - y*0.729);
                Square[2].x := Round(x+320) + width div 2;
                Square[2].y := Round(175 - y*0.729);
                Square[3].x := Round(x1+320) + width div 2;
                Square[3].y := Round(175 - y1*0.729);
                Square[4].x := Round(x1+320) - width div 2;
                Square[4].y := Round(175 - y1*0.729);
            end;
            FillPoly(4,Square);
            if level > 0 then
            begin
                turtle_theta := point(x,y,x1,y1);
                turn(left_angle);
                generate(x1,y1,Round(left_width_factor*width),
                    left_height_factor* left_height,
                    left_angle,level);
                turtle_theta := point(x,y,x1,y1);
                turn(-right_angle);
                generate(x1,y1,Round(right_width_factor*width),
                    right_height_factor* height,
                    right_angle,level);
            end;
        end;

begin
    write('Enter stem height: ');
    readln(height);
    write('Enter stem width: ');
    readln(width);
    write('Enter left alpha: ');
```

```
readln(left_alpha);
write('Enter right alpha: ');
readln(right_alpha);
write('Enter left branch angle: ');
readln(left_angle);
write('Enter right branch angle: ');
readln(right_angle);
write('Enter recursion level: ');
readln(level);
GraphDriver := 4;
GraphMode := EGAHi;
InitGraph(graphDriver,GraphMode,'');
left_width_factor := exp((-1n_2/left_alpha));
left_height_factor := exp((*1n_2/(3*left_alpha));
right_width_factor := exp((-1n_2/right_alpha);
right_height_factor := exp((-2*1n_2/(3*right_alpha)));
x := 0;
y := -235;
SetBkColor(9);
ClearDevice;
x1 := 0;
y1 := y + height;
SetColor(6);
SetFillStyle(1,6);
if Abs(x - x1) > Abs(y - y1) then
begin
    Square[1].x := Round(x+320);
    Square[1].y := Round(175 - y*0.729) + width div 2;
    Square[2].x := Round(x+320);
    Square[2].y := Round(175 - y*0.729) - width div 2;
    Square[3].x := Round(x1+320);
    Square[3].y := Round(175 - y1*0.729) - width div 2;
    Square[4].x := Round(x1+320);
    Square[4].y := Round(175 - y1*0.729) + width div 2;
end
else
begin
    Square[1].x := Round(x+320) - width div 2;
    Square[1].y := Round(175 - y*0.729);
    Square[2].x := Round(x+320) + width div 2;
    Square[2].y := Round(175 - y*0.729);
    Square[3].x := Round(x1+320) + width div 2;
    Square[3].y := Round(175 - y1*0.729);
    Square[4].x := Round(x1+320) - width div 2;
    Square[4].y := Round(175 - y1*0.729);
end;
FillPoly(4,Square);
turtle_theta := point(x,y,x1,y1);
turn(left_angle);
generate(x1,y1,Round(left_width_factor*width),
    left_height_factor*height, left_angle,level);
```

```
        turtle_theta := point(x,y,x1,y1);
        turn(-right_angle);
        generate(x1,y1,Round(right_width_factor*width),
            right_height_factor*height, right_angle, level);
        ch := ReadKey;
end.
```

The *generate* function begins by setting the turtle coordinates to the *x* and *y* coordinates passed to the function (which mark the beginning point for the function's operations. The *turtle_r* (step size) parameter is set to the height that was passed to the function. (The turtle angle was already set properly before the function was called.) The function makes the turtle step, extracts the new coordinates, sets the *width* parameter to the width passed to the function, decrements level, and then draws the branch.

Drawing such a branch was easy when I was using my C language graphics functions, since the line width in the *drawLine* function could be anything you wanted it to be. Unfortunately, Turbo Pascal has only thin and thick lines, so to do the same thing, we have to use the width to determine the bounds of a four-sided polygon for each branch, and then use the *FillPoly* procedure to draw the branch. If level has reached zero, this is the end; otherwise, the function turns the turtle by the left angle, appropriately scales down the length and width, and calls itself to make another left-hand line. When this has been done recursively, the turtle is reaimed to its original position when the function was called, rotated by the right angle, and generate is again called with the appropriately scaled parameters to do the right branch. Note that it is fairly easy to insert parameters into this program that will cause it to attempt to exceed the bounds of the display. Hopefully, Turbo Pascal will not allow you to come to any harm in this situation, but just the same, it is a good idea to avoid it.

Figure 10-2 is a chart of the parameters used to generate the trees that appear in Figures 10-3 through 10-11. Figure 10-3 shows three "stick" trees, each using the same parameters except for different values of a. They will give you an idea of how α affects the tree drawing. Figure 10-4 is set up to look as much like a real bare tree as possible. Figure 10-5 is similar, except that it uses a greater number of iterations to represent leaves. The program is set up to provide a blue sky-like background, and draw the lower-level branches brown and the upper-level (foliage) green. This gives a nice color display for Figure 10-5, but for the other trees shows some branches brown and some green. If this is objectionable, you may want to set the color statements for all the same color, as I did when I made the black and white figures.

Figure 10-6 is a one-sided curve that shows what happens when α is set close to zero. You must have a small value inserted or you will get a *divide by zero* error. Note that all of the displayed curve is generated very quickly and then a lot of time is spent computing nothing that will be displayed. Figure 10-7 makes use of the value of α that is supposed to be representative of the bronchial system. You can decide for yourself whether it is realistic (if you have ever seen a bronchial system) or determine how the parameters should be modified for a better representation. Figure 10-8 makes use of the value of α that is supposed to represent the arterial system. Figures 10-9 and 10-10 show the interesting curves that are obtained when the branching angle is set to ninety degrees. They don't look much like any real trees. Figure 10-11 is the same except that the branching angle is 85 degrees, which give a cockeyed tilt to the whole picture.

Figure 10-2. Parameters for Tree Drawing Program

Figure	Height	Width	Left a	Right a	Left Angle	Right Angle	Level
10-3a	100	1	1.1	1.1	25	25	6
10-3b	100	1	1.5	1.5	25	25	6
10-3c	100	1	2.0	2.0	25	25	6
10-4	120	20	2.0	2.2	24	26	6
10-5	80	20	2.0	2.2	20	28	14
10-6	200	35	2.0	0.00001	55	0	18
10-7	75	10	3.0	3.0	33	33	9
10-8	75	10	2.7	2.7	33	33	9
10-9	250	35	1.2	1.2	90	90	10
10-10	250	100	1.0	1.0	90	90	10
10-11	200	35	1.2	1.2	85	85	9

Figure 10-3. Stick Trees with Different α's

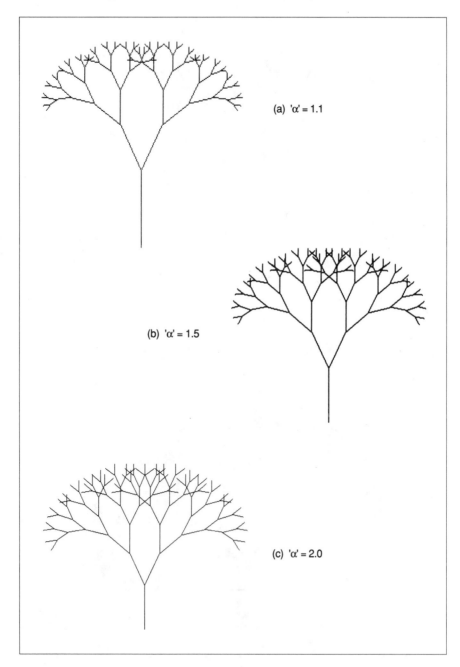

(a) 'α' = 1.1

(b) 'α' = 1.5

(c) 'α' = 2.0

Figure 10-4. Bare Tree

Figure 10-5. Tree with Foliage

Figure 10-6. One-Sided Tree

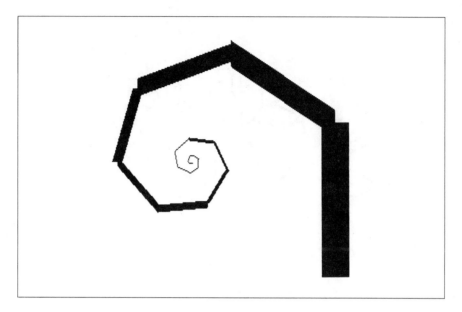

Figure 10-7. Bronchial System Tree

Figure 10-8. Arterial System Tree

Figure 10-9. Tree with Ninety Degree Branch Angles

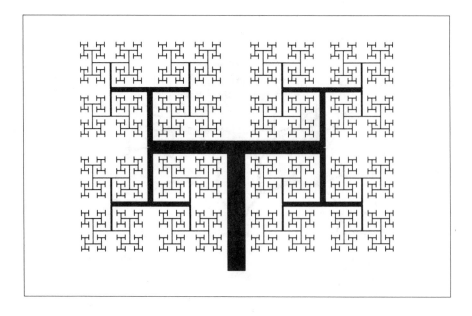

Figure 10-10. Tree with 90 Degree Branch Angles and Wider Stem

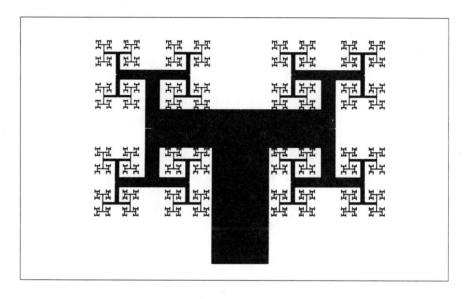

Figure 10-11. Tree with 85 Degree Branch Angles

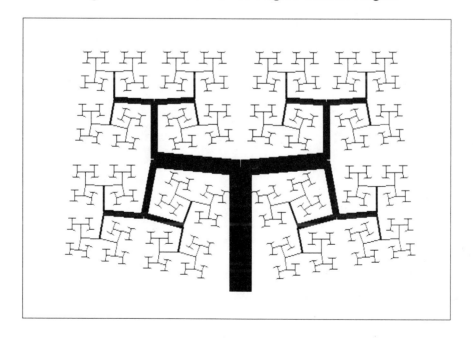

11

Working with Circles

So far, we have done all of our generating of fractals using straight line segments. Since these simple geometric figures have resulted in very complex fractal curves that are of great beauty and interest, we might suspect that we can use circles—which are geometric figures of a greater order of complexity—to generate fractals that are even more interesting.

To some extent this is true. The problem is that the geometry required to put together a fractal pattern of circles, determining their radii and their coordinates, is so much more complicated than that used for fractal lines, that it has barely been explored. And those who have investigated it, although they are more than willing to display their pretty pictures, are not willing to reveal the secrets of how they compute their fractal curves.

So we'll do the best we can, and in this chapter show how to draw two representative fractals using circles. In the course of this, you will observe that we have to make use of some geometric formulas that rarely occur in geometry books, and when they do, are not always clearly explained.

With this introduction, you are on your own. The field of circle fractals is a big one with lots of room for discovery, but you may have to do a lot more research into geometry than you thought you might ever want to do.

Apollonian Packing of Circles

This is one of the most basic of fractal curves involving circles. First we draw three circles, each of which is externally tangent to the other two. The result of their joining is a curvilinear triangle. Next, we draw a circle that will fit into the curvilinear triangle, being tangent to each of the three given circles. This yields three smaller, curvilinear triangles.

In a similar manner, we draw a circle in each of these triangles, tangent to the three circles that make up the triangle. Now, from each curvilinear triangle, three more smaller curvilinear triangles are generated. Continuing the process infinitely yields the ultimate fractal curve, but for practical programming, we will stop the process after six or seven iterations; before too long, the circles become so small that they either appear on our screen as dots, or not at all.

It can be shown that the resulting curve is fractal and has a Hausdorff-Besicovitch dimension (discovered by Boyd) of approximately 1.3058. This begs the question of how we perform this repeated circle drawing task. Provided we can figure out how to draw a circle within a curvilinear triangle, knowing the coordinates of the center and the radius of each of the three circles that generate the curvilinear triangle, then we can use a recursive process similar to what we have used for previous fractals to draw repeated circles. It is the former that is the problem, however. Fortunately, by using the principles of analytic geometry, we can determine the x and y coordinates of the new circle, providing we know its radius. The process is tedious but not difficult.

Soddy's Formula

It turns out that there is a formula for the radius of the circle inscribed within the curvilinear triangle, given solely in terms of the radii of the three circles that make up the triangle. It is called *Soddy's formula*. One form of this formula is the expression:

```
2(1/a² + 1/b² + 1/c² + 1/r²) = (1/a + 1/b + 1/c + 1/r)²   (Equation 11-1)
```

where *a*, *b*, and *c* are the radii of the three given tangent circles and *r* is the radius of the circle that is to be drawn tangent to the three given circles. The form of the expression that we will work with is:

```
1/r = 1/a + 1/b + 1/c + 2 (1/bc + 1/ac + 1/ab)        (Equation 11-2)
```

Once we know the radius of the circle to be drawn and the radii and center coordinates of the three given circles, we can do some complicated but very straightforward mathematics to determine the *x* and *y* coordinates of the center of the circle to be drawn. We can then draw the circle.

Program for Apollonian Circle Packing

Figure 11-1 lists the program for the Apollonian packing of circles. It first asks for a level to which recursion will continue before the program terminates. Then the three original circles are drawn. If you wish to change the location of these circles, you can either calculate their locations or arrange them by running just the first part of the program several times and changing the circle coordinates until you find the desired location and conditions of tangency by trial and error.

Figure 11-1. Program to Perform Apollonian Packing of Circles

```
program apollo;

uses CRT,Graph,Fractal;
```

```
var
    color,graphDriver,GraphMode,i,generator_size,level,
        init_size: integer; ch: char;
    a,b,c,s,cs,bs,xa,xb,xc,ya,yb,yc,xs,ys,temp: real;

procedure node(xa: real; ya: real; a: real; xb: real; yb:
    real; b: real; xc: real; yc: real; c: real; xs: real;
    ys: real; s: real; level: integer); forward;

procedure gen_circle(xa: real; ya: real; a: real; xb: real;
    yb: real; b: real; xc: real; yc: real; c: real; level:
    integer);

    var
        s,temp: real;

    begin
        dec(level);
        s := 1/a + 1/b + 1/c + 2*(sqrt(1/(b*c) + 1/(c*a) +
            1/(a*b)));
        s := 1/s;
        temp := (s+a)*(s+a) - (s+b)*(s+b) - xa*xa + xb*xb
            - ya*ya + yb*yb;
        ys := (temp*(xc-xa) - (xb - xa)*((s+a)*(s+a) -
            (s+c) *(s+c) - xa*xa + xc*xc - ya*ya +
            yc*yc))/(2*((yb-ya)*(xc-xa) - (yc-ya)*
            (xb - xa)));
            xs := (temp - 2*ys*(yb - ya))/(2*(xb - xa));
        inc(color);
        SetFillStyle(1,color);
        SetColor(color);
        FillEllipse(Round(xs+320),Round(175 -
            ys*0.7291666), Round(s),Round(s*0.7291666));
        if level > 0 then
            node(xa,ya,a,xb,yb,b,xc,yc,c,xs,ys,s,level);
    end;

procedure node(xa: real; ya: real; a: real; xb: real; yb:
    real; b: real; xc: real; yc: real; c: real; xs: real;
    ys: real; s: real; level: integer);
    begin
        gen_circle(xa,ya,a,xb,yb,b,xs,ys,s,level);
        gen_circle(xb,yb,b,xc,yc,c,xs,ys,s,level);
        gen_circle(xa,ya,a,xc,yc,c,xs,ys,s,level);
    end;

begin

    write('Enter level (1 - 8): ');
    readln(level);
```

```
  if level < 1 then
      level := 1;
  GraphDriver := 4;
  GraphMode := EGAHi;
  InitGraph(graphDriver,GraphMode,'');
  a := 625;
  b := 375;
  c := 945;
  color := 1;
  xa := -725;
  ya := 235;
  xb := 275;
  yb := 268;
  xc := 180;
  yc := -1048;
  SetFillStyle(1,color);
  setColor(color);
      FillEllipse(Round(xa+320),Round(175
          -ya*0.7291666),Round(a),Round(a*0.7291666));
  inc(color);
  SetFillStyle(1,color);
  SetColor(color);
  FillEllipse(Round(xb+320),Round(175
      -yb*0.7291666),Round(b),Round(b*0.7291666));
  inc(color);
  SetFillStyle(1,color);
  SetColor(color);
  FillEllipse(Round(xc+320),Round(175
      -yc*0.7291666),Round(c),Round(c*0.7291666));
  gen_circle(xa,ya,a,xb,yb,b,xc,yc,c,level);
  ch := ReadKey;
end.
```

Next, the *gen_circle* function is run to draw and fill the new circle in the first (biggest) curvilinear triangle. It starts by using Soddy's formula to determine the radius of the circle to be drawn. Next, it computes the x and y coordinates of the center for the new circle. Then the circle is filled.

The color increments each time a circle is drawn, so that there will be a distinction between the different circles. Note that there are two alternate (and redundant) methods for computing the coordinates of the center of the circle (*xs* and *ys*). Unfortunately, if the x coordinates of the centers of two of the three circles making up the curvilinear triangle are the same, using one of these methods will cause a divide by zero, resulting in the program crashing. Consequently, if one pair of *x* coordinates are equal, the other pair is used. Observe that all three *x* coordinates cannot be the same; if they were, the three circles would be in a straight line and no curvilinear triangle would be generated.

Once the *x* and *y* coordinates of the center of the new circle and its radius are known, the *FillEllipse* function can then be used to fill the new circle. Each time *gen_circle* is called, it reduces the level by one from that transferred by the calling function. When the level is reduced to zero, the function terminates after filling the new circle; otherwise, the function *node* is called to compute and fill circles in each of the three new curvilinear triangles comprising two of the original three circles and the new circle. Plate 3 (in the color section) shows the figure produced by running this program.

Inversion

In producing fractal curves involving circles, there is an interesting mapping technique that can transform a fairly simple pattern of circles into a much more interesting one. The technique also has uses in geometry, where it can often simplify complex relationships so that proofs of theorems become much simpler. The technique is called *inversion*. It makes use of a given circle to map all of the points on a plane onto the plane, except for the point at the very center of the circle. The mapping is done as follows:

1. A line is drawn from the center of the circle, O, to the point to be mapped, P.

2. The new mapped point, P', is placed on the line OP so that the product of the distances OP and OP' is equal to the square of the radius of the circle, r^2.

Inversion has several interesting properties. They include:

1. Any circle whose circumference passes through the center of the reference circle, *O*, maps into a straight line parallel to a tangent to the circle being mapped at *O*.

2. Any circle that is orthogonal to the circle of inversion inverts into itself.

3. Any other circle maps into a circle.

4. You might suspect that if a circle maps into another circle, the center of the first circle would map into the center of the new circle, but this is not true. If you want to perform an inversion and then plot it on the screen, you cannot successfully use data from the original circle to determine the center and radius of the mapped circle and then draw it using this information. You must map every point on the circle to its new location and plot it there.

In order to perform the inversions in a real computer program, we make use of the procedure *inverseOval*, which is listed as part of Figure 11-2. This program is one that could be used to replace the Turbo Pascal procedure for drawing circles, but it has the additional property that it performs an inversion on every point of the circle before plotting it to the screen.

Figure 11-2. Program to Generate Pharaoh's Breastplate

```
program pharaoh;

uses CRT,Graph,Fractal;

var
    graphDriver,GraphMode,i,generator_size,level,init_size:
        integer;
    ch: char;
    a_line,b_line,x_o,y_o,radius,r_sq,height: double;
    xbig,ybig,rbig,xtan,ytan,rtan: real;
```

```
procedure inverseOval(x: real; y: real; b: real; color:
    integer; aspect: real);

    var
        i, bnew,new_col, new_row: integer;
        length,new_length: real;
        a,a_square, b_square, two_a_square, two_b_square,
        four_a_square, four_b_square,d,row,col: longint;

    begin
        a := Round(b/aspect);
         b_square := Round(b*b);
        a_square := (a*a);
        row := Round(b);
        col := 0;
        two_a_square := a_square shl 1;
        four_a_square := a_square shl 2;
        four_b_square := b_square shl 2;
        two_b_square := b_square shl 1;
        d := two_a_square * ((row  -1)*(row )) + a_square
            + two_b_square*(1-a_square);
        while a_square*row  > b_square * col do
        begin
            length := sqrt((x_o - col - x)*(x_o-col - x)
                + (y_o - row - y)*(y_o -row - y));
            new_length := r_sq/length;
            new_col := Round(x_o - (x_o - col -
                x)*new_length/length);
            new_row := Round(-y_o + (y_o - row -
                y)*new_length/length);
            PutPixel(new_col+320,175 -
                Round(new_row*0.729),color);
            length := sqrt((x_o + col - x)*(x_o + col -
                x) + (y_o - row - y)*(y_o - row - y));
            new_length := r_sq/length;
            new_col := Round(x_o - (x_o + col -
                x)*new_length/length);
            new_row := Round(-y_o + (y_o - row -
                y)*new_length/length);
            PutPixel(new_col+320,175 -
                Round(new_row*0.729),color);
            length := sqrt((x_o - col - x)*(x_o - col -
                x) + (y_o + row - y)*(y_o + row - y));
            new_length := r_sq/length;
            new_col := Round(x_o - (x_o - col -
                x)*new_length/length);
            new_row := Round(-y_o + (y_o + row -
                y)*new_length/length);
            PutPixel(new_col+320,175 -
                Round(new_row*0.729),color);
            length := sqrt((x_o + col - x)*(x_o + col -
```

```
        x) + (y_o + row - y)*(y_o + row - y));
    new_length := r_sq/length;
    new_col := Round(x_o - (x_o + col -
        x)*new_length/length);
    new_row := Round(-y_o + (y_o + row -
        y)*new_length/length);
    PutPixel(new_col+320,175 -
        Round(new_row*0.729),color);
    if d >= 0 then
    begin
        dec(row);
        d := d - four_a_square*(row);
    end;
    d := d + two_b_square*(3 + (col shl 1));
    inc(col);
end;

d := two_b_square * (col + 1)*col +
    two_a_square*(row * (row  -2) +1) +
    (1-two_a_square)*b_square;
while row + 1 <> 0 do
begin
    length := sqrt((x_o - col - x)*(x_o - col -
        x) + (y_o - row - y)*(y_o - row - y));
    new_length := r_sq/length;
    new_col := Round(x_o - (x_o - col -
        x)*new_length/length);
    new_row := Round(-y_o + (y_o - row -
        y)*new_length/length);
    PutPixel(new_col+320,175 -
        Round(new_row*0.729),color);
    length := sqrt((x_o + col - x)*(x_o + col -
        x) + (y_o - row - y)*(y_o - row - y));
    new_length := r_sq/length;
    new_col := Round(x_o - (x_o + col -
        x)*new_length/length);
    new_row := Round(-y_o + (y_o - row -
        y)*new_length/length);
    PutPixel(new_col+320,175 -
        Round(new_row*0.729),color);
    length := sqrt((x_o - col - x)*(x_o - col -
        x) + (y_o + row - y)*(y_o + row - y));
    new_length := r_sq/length;
    new_col := Round(x_o - (x_o - col -
        x)*new_length/length);
    new_row := Round(-y_o + (y_o + row -
        y)*new_length/length);
    PutPixel(new_col+320,175 -
        Round(new_row*0.729),color);
    length := sqrt((x_o + col - x)*(x_o + col -
        x) + (y_o + row - y)*(y_o + row - y));
```

```
            new_length := r_sq/length;
            new_col := Round(x_o - (x_o + col -
                x)*new_length/length);
            new_row := Round(-y_o + (y_o + row -
                y)*new_length/length);
            PutPixel(new_col+320,175 -
                Round(new_row*0.729),color);
            if d <= 0 then
            begin
                inc(col);
                d := d + four_b_square*col;
            end;
            dec(row);
            d := d + two_a_square * (3 - (row shl 1));
        end;
        b := b + 1;
    end;

procedure gen_circle(x: real; y: real; radius: real);
    begin
        inverseOval(x,y,radius,15,1.0);
        inverseOval(-x,y,radius,15,1.0);
    end;

begin
    GraphDriver := 4;
    GraphMode := EGAHi;
    InitGraph(graphDriver,GraphMode,'');
    r_sq := 400000;
    xbig := 0;
    ybig := 0;
    rbig := 220;
    rtan := 140;
    xtan := 0;
    ytan := ybig + rbig - rtan;
    y_o := ybig - rbig;
    x_o := xbig;
    Ellipse(Round(xbig+320),175 -
        Round(ybig*0.729),0,360,Round(rbig),
        Round(0.729*rbig));
    Ellipse(Round(xtan+320),175 -
        Round(ytan*0.729),0,360,Round(rtan),
        Round(0.729*rtan));
    a_line := r_sq/(2*rbig);
    b_line := r_sq/(2*rtan);
    height := (b_line - a_line);
    radius := height/2;
    height := radius*sqrt(2.0);
    for i:=0 to 15 do
    begin
        gen_circle(x_o + height*i,y_o + a_line +
```

```
        radius,radius);
    gen_circle(x_o + height*i,y_o + a_line +
        radius/2,radius/2);
    gen_circle(x_o + height*i,y_o + b_line -
        radius/2,radius/2);
    gen_circle(x_o + height*i + height/2,y_o + a_line
        + 3*radius/4,radius/4);
    gen_circle(x_o + height*i + height/2,y_o + b_line
        - 3*radius/4,radius/4);
    gen_circle(x_o + height*i + height/2,y_o + a_line
        + radius/8, radius/8);
    gen_circle(x_o + height*i + height/2,y_o + b_line
        - radius/8, radius/8);
    gen_circle(x_o + height*i + height/2,y_o + a_line
        + 5*radius/12,radius/12);
    gen_circle(x_o + height*i + height/2,y_o + b_line
        - 5*radius/12,radius/12);
    gen_circle(x_o + height*i + 0.4*height,y_o +
        a_line + 0.3*radius,radius/10);
    gen_circle(x_o + height*i + 0.6*height,y_o +
        a_line + 0.3*radius,radius/10);
    gen_circle(x_o + height*i + 0.4*height,y_o +
        b_line - 0.3*radius,radius/10);
    gen_circle(x_o + height*i + 0.6*height,y_o +
        b_line - 0.3*radius,radius/10);
    end;
    ch := ReadKey;
end.
```

Pharaoh's Breastplate

Pharaoh's Breastplate is the name given by Mandelbrot to a figure created using inversion. The figure that we are going to generate here is not quite the same as Mandelbrot's, but it does make use of inversion and gives you some idea of how the technique works. Figure 11-3 shows a pattern of circles, before inversion.

The two large circles represent the final reference circles. They are mapped by inversion into the upper and lower horizontal lines. The other circles are tangent to each other and to the horizontal lines. They can extend as far to the left and right as you want them to. The number of circles is set by a parameter in a *for* loop. For Figure 11-3, the parameter is set to 3; when we perform the final inversion program, the parameter is set to 15.

A word needs to be said about the number that is inserted into the program for *r_sq*. This is the square of the radius of the reference circle that is used in the inversion mapping. The center of this circle is always at the point of tangency of the two large circles, so that they are mapped into straight lines by the inversion. The radius of the reference circle determines where these two lines are located. All of the other circles that the program generates are referenced to these two lines. When the inversion of these circles takes place, they end up being referenced to the two large circles, so that no matter what value is used for the radius of the reference circle, all of the smaller circles will be referenced to the same place in the final drawing.

For Figure 11-3, where we are actually showing the uninverted small circles, we chose a value for *r_sq* that would cause the two reference straight lines to appear at a convenient location on the display. There is another consideration, however. The smaller the reference circle radius, the closer together will be the two straight lines that represent the mapping of the two large circles. Consequently, all of the small circles will be smaller, and when we run *inverseOval*, there will be fewer points generated to make up these small circles. If we generate a lot fewer points than we would normally have drawn to produce the inverted circle, the resulting circle will be rather coarse and tend to lose its circular shape. Consequently, we want to make *r_sq* large, but not so large that we exceed the range of values that the computer can conveniently handle.

The value of 400000 used in the pharaoh program is a good compromise for this particular program. In both the program that created Figure 11-3 and the final program, we start by generating the two reference circles and then invert them to create the two parallel lines. (In the final program, the lines are not drawn.) Then we generate all of the necessary circles in reference to the two parallel lines. The mathematics is complex, but straightforward; if you wish, you can tackle the geometry for determining the radius and center coordinates of each set of circles to see how the results given in the program are obtained. It may be that there are some neat tricks of inversion or recursion that I have missed that will simplify the program, but the approach used here is the simplest one and gets down to a reasonably fine detail of circles. If you want to add more circles, go right ahead and compute the necessary radii and center coordinates.

Figure 11-3. Pharaoh's Breastplate Circles before Inversion

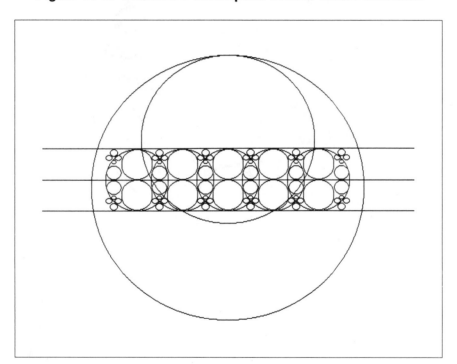

One final word: Turbo Pascal has a procedure called *Circle* that can be used to draw circles. Unfortunately, the radius of the circle drawn is not directly related to either the *x* or *y* coordinates, but depends upon the setting of the aspect ratio between them. To avoid introducing complications involved in determining the correct aspect ratio, we make use of the Turbo Pascal procedure *Ellipse* to draw our large circles and supply our own coordinate conversion. The program listing in its entirety is shown in Figure 11-2. The resulting picture of Pharaoh's breastplate is shown in Figure 11-4.

Figure 11-4. Pharaoh's Breastplate

12

The Mandelbrot Set

The Mandelbrot set is probably the most well-known of the fractal curves. In almost every magazine, you will come across an article on the Mandelbrot set and some examples of the pictures of its displays. Almost every bulletin board has a Mandelbrot set program. Originally, the Mandelbrot set was discovered by Benoit Mandelbrot when he was investigating the behavior of the iterated function:

$$z_{n+1} = z_n^2 + c \qquad \text{(Equation 12-1)}$$

where both z and c are complex numbers. First, to get a feel for the function, consider the very simple situation where z_0 is a real number and c is zero. If z_0 is 1, the value of z remains at 1, no matter how many iterations are performed. If z_0 is less than 1, the function z_n will approach zero as n approaches infinity. If z_0 is greater than one, z_n will approach infinity. The speed at which z_n approaches zero for numbers less than one, or approaches infinity for numbers greater than one, depends upon the original value of the function, z_0. The smaller this value, the faster the function will approach zero for starting values less than one. The larger the value, the faster the function will approach infinity for starting values greater than one.

This over-simplified example is fairly easy to understand. When z_0 becomes a complex number, and/or c becomes a complex number instead of zero, the situation becomes much more complex. In fact, for years mathematicians steered away from the complexities of this kind of expression, assuming that it eventually reached limiting values in some fairly regular fashion. It was only when computers were applied to the problem extensively that it was discovered that the behavior of the expressions was quite chaotic, and that the result of performing many iterations of the expression for various values yielded fractal curves.

In plotting the function in some meaningful way, we want to show what happens to the expression for some range of reasonable initial conditions. To do this, we shall perform a sufficient number of iterations to determine the behavior of the iterated function. The attractor for the Mandelbrot set is infinity. It can be shown that if the magnitude of the function ever exceeds two, it will eventually be attracted to infinity. For practical purposes, we find that for most cases, if the magnitude of the function does not exceed two within 512 iterations, it probably never will. We will color such points black. The rest of the colors that we have available will be used to indicate the speed with which the function is approaching infinity, based upon the number of iterations that are required for the magnitude to become greater than two. What we have done is to cycle through the sixteen colors available to us on the EGA, incrementing the color value once for each iteration. Thus, there is a color change each time the number of iterations required to blow up increases by one. However, what you do with the colors to indicate various numbers of iterations is pretty much up to your own imagination. You can cycle as has been done here, or with groups of iterations assigned to different color values, and you can assign colors any way you wish. The result is often not only artistic, but reveals a different meaning about how the function behaves as the parameters are changed.

Once we know what we are going to do with the colors, we have to decide what the x and y coordinates of our mapping of the function will represent. There are two basic ways to go here. One is to let z_0 be equal to zero and let the x and y coordinates of the display represent the real and imaginary parts of c as they change over some selected range of values. This gives rise to the well known Mandelbrot set. The other approach is to select a value for c and let the x and y axis be equal to the real and imaginary parts of z_0 as it changes over a selected range of values. This gives rise to the Julia sets, which will be described in Chapter 13. Interestingly enough, the Mandelbrot set forms a sort of map of the Julia sets; if you select a point on the Mandelbrot set and enlarge the area around it sufficiently, you get a pattern that is very similar to the Julia set for that same point. Thus, you can use the Mandelbrot set to determine which Julia sets you would like to plot.

Simplified Mandelbrot Set Program

Figure 12-1 lists a plain vanilla program for generating the Mandelbrot set. The heart of the algorithm consists of two *for* loops and one *while* loop. The *for* loops make sure that each pixel in the vertical and horizontal directions is assigned a color value by the algorithm. The *while* loop performs the successive iterations of the equations until the magnitude of the square of the function is larger than 4 or until the number of iterations reaches 512, whichever occurs first. In squaring the complex function z, and adding the complex number c, the real part is:

$$x_n = x_{n-1}^2 - y_{n-1}^2 + p \qquad \text{(Equation 12-2)}$$

and the imaginary part is:

$$y_n = 2x_{n-1}y_{n-1} + q \qquad \text{(Equation 12-3)}$$

In determining how to optimize the performance of our algorithm, we need to be aware of the fact that the *while* loop can be iterated as many as 1.147×10^8 times, the *for* loop for rows is iterated 224,000 times, whereas the *for* loop for columns is called only 640 times. Thus, it is most important to minimize the time spent in calculations in the *while* loop, and fairly important to minimize the time spent in the *for* loop for rows, while the time spent in the *for* loop for columns is relatively insignificant. We have minimized time spent in the *while* loop by using the square of the magnitude $(x^2 + y^2)$ for the comparison, so that we don't have to perform any square roots. In addition, in computing y we have avoided multiplying by two, substituting an addition, which is a much faster process.

Figure 12-1. Program to Generate Mandelbrot Set

```
program mandel;

uses CRT,Graph;

const
     maxcol = 640;
     maxrow = 350;
     max_colors = 16;
     max_iterations = 512;
     max_size = 4;

var
    Q: array[0..349] of real;
    XMax,YMax,XMin,YMin: real;
    P,deltaP, deltaQ, X, Y, Xsquare, Ysquare: real;
    i,color, row, col,GraphDriver,GraphMode: integer;
    ch: char;

begin
     XMax := 1.2;
     XMin := -2.0;
     YMax := 1.2;
     YMin := -1.2;
     GraphDriver:= 4;
     GraphMode := EGAHi;
     InitGraph(GraphDriver,GraphMode,'');
     deltaP := (XMax - XMin)/(maxcol);
     deltaQ := (YMax - YMin)/(maxrow);
     Q[0] := YMax;
     for row:=1 to maxrow do
         Q[row] := Q[row-1] - deltaQ;
     P := XMin;
     for col:=0 to maxcol do
     begin
          if KeyPressed then
       exit;
         for row:=0 to maxrow do
         begin
              X := 0.0;
              Y := 0.0;
              Xsquare := 0.0;
              Ysquare := 0.0;
              color := 1;
              repeat
```

```
            Xsquare := X*X;
            Ysquare := Y*Y;
            Y := 2*X*Y + Q[row];
            X := Xsquare - Ysquare + P;
            inc(color);
        until (color>=max_iterations) or (Xsquare +
            Ysquare >= max_size);
        PutPixel(col, row, (color mod max_colors));
        end;
        P := P + deltaP;
    end;
    ch := ReadKey;
end.
```

In the row *for* loop, we have avoided computing the value of q at each pass through the loop. This would require a total of 224,000 calculations of q during passes through the loop, although there are only 350 distinct values of q used. Consequently, we calculate these 350 values first and put them in an array and then select the proper one for each pass through the loop. The program includes a test of the keyboard with provision to exit when a key is struck, in case the user doesn't want to complete a picture. This test, however, is done within the outermost for loop. This means that, if a key is struck, the program will complete the column that it is working on before exiting. This delays the exit a little, but prevents the key check from slowing up the program while a picture is being drawn. Plate 4 (in the color section) is a picture of the entire Mandelbrot set as drawn by this program. The letters and arrows on this figure have been added to show the locations where various parameters for the Julia sets of the next chapter have been located. Various expansions of the Mandelbrot set are shown in Plates 5, 6, 7, 8, and 9 (in the color section).

Improved Mandelbrot Set Program

The program described above produces the Mandelbrot set, but it doesn't have any bells and whistles to make our job easier. An improved program would have the following characteristics:

1. It should be capable of saving a partial picture at any time we want to interrupt it and then picking up from where it left off to complete the picture at some future time.

2. It should be able to save a completed picture and redisplay it on the screen in just a few seconds.

3. It should be capable of selecting a small section of a picture and blowing it up to a full-screen display.

4. It should provide some more sophisticated coloring capabilities.

5. It should be as user-friendly as possible.

The program listed in Figure 12-2 will meet all of the requirements for the improved program. Unfortunately, it and the programs given in the next three chapters are beginning to get somewhat complicated.

One of the aims of this book is to keep all of the programs simple enough so that you can read the listings and understand what's going on. These programs tend to violate that goal. Please note, however, that much of the complexity occurs in creating a set of user-friendly menus which can be used to easily set the program parameters. You can eliminate the associated procedures at the beginning and just set in some fixed values to get going. Then, when the main part of the program is working, you can add the menu portions. Note that it is possible to make a main program (for testing) that only calls the menu programs, and possibly prints out the parameters that they return. Alternately, you can use the entire Mandelbrot set program, but comment out the portion that generates the Mandelbrot set. In this way you can get the menus going and debugged without having to generate a Mandelbrot set at every trial.

Figure 12-2. Improved Mandelbrot Set Program

```
program cmandel;

uses CRT,Graph,Fractal;

const
    maxcol: integer = 639;
    maxrow: integer = 349;
    max_size: real = 4;
    file_string: array[0..1] of string[7] =
        (('mandel'),('mpalet'));
    file2: string[5] = '.pcx';
    file3: string[5] = '.pal';
    rect: array[1..4] of PointType = ((x: 0; y: 0),
                                      (x: 639; y: 0),
                                      (x: 639; y: 349),
                                      (x: 0; y: 349));

var
    graphDriver,GraphMode,i,j,key,generator_size,level,
        init_size,file_no,color,row,col,error,start_col,
        end_mask, m,color_option,max_iterations: integer;
    deltaP,deltaQ,X,Y,Xsquare,Ysquare,P: real;
    ch1: char;
    colors: array[0..14,0..2] of integer;
    file1: string[3];
    file_name,file_name2: string[12];
    f: file of byte;
    f_pal: file of integer;
    Q: array[0..349] of real;

function menu(types: integer): integer;

const
    TextData: array[0..3] of string[45] =
        (('Generate the Mandelbrot Set'),
         ('Expand a Section of an Existing Mandelbrot Set'),
         ('Complete an Unfinished Mandelbrot Set'),
         ('Quit'));
    TextData2: array[0..1] of string[45] =
        (('Select colors and ranges'),
         ('Cycle through colors in order'));

var
    i,k: integer;
    ch1: char;
begin
    k := 0;
    Window(1,1,80,25);
    TextBackground(0);
    clrscr;
    if types = 0 then
    begin
        Window(10,6,59,13);
```

(handwritten annotations: "cmandel2", "double extended", "double extended")

```
        m := 3;
    end
    else
    begin
        Window(10,6,59,11);
        m := 1;
    end;
    TextBackground(1);
    clrscr;
    GotoXY(3,2);
    TextColor(10);
    write('Make your choice with the up and down arrows:');
    repeat
        for i:=0 to m do
        begin
            if i = k then
            begin
                TextColor(1);
                TextBackground(15);
            end
            else
            begin
                TextColor(15);
                TextBackground(1);
            end;
            GotoXY(3,i+4);
            if types = 0 then
                write(TextData[i])
            else
                write(TextData2[i]);
        end;
        ch1:= ReadKey;
        if ch1 = char($00) then
        begin
            ch1 := ReadKey;
            case ch1 of
                'P': begin
                        if k = 3 then
                            k := 0
                        else
                            inc(k);
                    end;
                'H': begin
                        if k = 0 then
                            k := 3
                        else
                            dec(k);
                    end;
            end;
        end;
    until ch1 = char($0D);
    window(1,1,80,25);
    menu := k;
end;
```

```
function set_colors: integer;

const
    color_set: array[0..1] of integer = (1,1);

var
    i,k: integer;
    ch1: char;
begin
    TextBackground(4);
    clrscr;
    k := 0;
    Window(10,4,70,22);
    TextBackground(0);
    clrscr;
    GotoXY(3,1);
    TextColor(10);
    writeln('Enter Upper Limit Number and use Arrows to Set
        Colors:');
    writeln;
    TextColor(14);
    write('  Start Iters    End Iters    Color #1  Color #2');
     writeln;
    i := 0;
    colors[0,0] := 0;
    colors[1,0] := 0;
    while colors[i,0] < max_iterations do
    begin
        inc(i);
        TextColor(15);
        GotoXY(6,i+5);
        write(colors[i-1,0]);
        GotoXY(20,i+5);
        Read(colors[i,0]);
        if colors[i,0] <= colors[i-1,0] then
        begin
            colors[i,0] := colors[i-1,0] + 1;
            GotoXY(20,i+5);
            write(colors[i,0]);
        end;
        if (i = 13) or (colors[i,0] > max_iterations)
        then
        begin
            colors[i,0] := max_iterations;
            GotoXY(20,i+5);
            write(colors[i,0]);
        end;
        GotoXY(30,i+5);
        TextColor(color_set[0]);
        write(char($DB),char($DB),char($DB),char($DB),
            char($DB),char($DB));
        TextColor(color_set[1]);
        GotoXY(40,i+5);
        write(char($DB),char($DB),char($DB),char($DB),
            char($DB),char($DB));
        for k:= 0 to 1 do
```

```
        begin
            GotoXY(36+ 10*k,i+5);
            repeat
                ch1:= ReadKey;
                if ch1 = char($00) then
                begin
                    ch1 := ReadKey;
                    case ch1 of
                        'P': begin
                            if color_set[k] = 15 then
                                color_set[k] := 0
                            else
                                inc(color_set[k]);
                            end;
                        'H': begin
                            if color_set[k] = 0 then
                                color_set[k] :=  15
                            else
                                dec(color_set [k]);
                            end;
                    end;
                end;
                GotoXY(30+10*k,i+5);
                TextColor(color_set[k]);
                write(char($DB),char($DB),char($DB),
                    char($DB),char($DB),char($DB));
            until ch1 = char($0D);
        end;
        colors[i,1] := color_set[0];
        colors[i,2] := color_set[1];
        set_colors := i;        end;
    Window(1,1,80,25);
    clrscr;
    TextColor(15);
end;

begin
    DirectVideo := false;
    GraphDriver := 4;
    GraphMode := EGAHi;
    repeat
        key := menu(0);
        if key = 3 then
            exit;
        if key < 2 then
        begin
            TextBackground(0);
            clrscr;
            Window(15,8,66,11);
            TextBackground(4);
            clrscr;
            GotoXY(3,2);
            TextColor(15);
            write('Enter number of iterations desired
                (16-512): ');
            readln(max_iterations);
```

```
    if max_iterations < 16 then
        max_iterations := 16;
    if max_iterations > 512 then
        max_iterations := 512;
end;
color_option := menu(1);
if (key < 2) and (color_option = 0) then
    end_mask :=  set_colors;
TextBackground(0);
clrscr;
Window(15,8,66,11);
TextBackground(2);
clrscr;
GotoXY(3,2);
TextColor(15);
if key <> 0 then
begin
    write('Enter Mandelbrot Set number (00 - 99):     ');
    readln(file_no);
    Str(file_no,file1);
    Window(1,1,80,25);
    if file_no < 10 then
        file_name := Concat
            (file_string[0],'0',file1,file2)
    else
        file_name := Concat
            (file_string[0],file1,file2);
    error := restore_screen(file_name);
    if error = 0 then
        exit
end;
CURSOR_X := 0;
CURSOR_Y := 0;
case key of
0: begin
    start_col := 0;
    XMax := 1.2;
    XMin := -2.0;
    YMax := 1.2;
    YMin := -1.2;
    file_name := Concat
        (file_string[0],'00',file2);
    InitGraph(GraphDriver,GraphMode,'');
    SetFillStyle(1,7);
    FillPoly(4,rect);
    end;
2: begin
    if (error = 639) or (error = 0) then
        exit;
    start_col := 8 * (error div 8);
    Assign(f,file_name);
    Erase(f);
    if color_option = 0 then
    begin
    if file_no < 10 then
        file_name := Concat
```

```
                    (file_string[1],'0',file1,file3)
        else
            file_name := Concat
                (file_string[1],file1,file3);
            Assign(f_pal,file_name2);
            Reset(f_pal);
            for i:= 0 to 14 do
                for j:= 0 to 2 do
                    read(f_pal,colors[i,j]);
            read(f_pal,max_iterations);
            read(f_pal,end_mask);
            Close(f_pal);
        end;
    end;
    1: begin
            move_cursor(0,15,0,0);
            move_cursor(1,15,CURSOR_X,CURSOR_Y);
            XMax := TXMax;
            XMin := TXMin;
            YMax := TYMax;
            YMin := TYMin;
            start_col := 0;
            SetFillStyle(1,7);
            FillPoly(4,rect);
        end;
    end;
deltaP := (XMax - XMin)/(maxcol);
deltaQ := (YMax - YMin)/(maxrow);
Q[0] := YMax;
for row:=1 to maxrow do
    Q[row] := Q[row-1] - deltaQ;
    P := XMin + start_col * deltaP;
for col:=start_col to maxcol do
begin
    if KeyPressed then
    begin
        file1 :=
            save_screen(0,0,col,349,file_name);
        gotoxy(1,24);
        file_name := Concat
            (file_string[0],file1,file2);
        write('File name is: ',file_name);
        if color_option = 0 then
        begin
            file_name := Concat
                (file_string[1],file1,file3);
            Assign(f_pal,file_name);
            Rewrite(f_pal);
            for i:= 0 to 14 do
                for j:= 0 to 2 do
                    write(f_pal,colors[i,j]);
            write(f_pal,max_iterations);
            write(f_pal,end_mask);
            Close(f_pal);
        end;
        exit;
```

```
        end;
        for row:=0 to maxrow do
        begin
            X := 0;
            Y := 0;
            Xsquare := 0;
            Ysquare := 0;
            color := 1;
            while (color<max_iterations) and
                ((Xsquare + Ysquare) < max_size) do
            begin
                Xsquare := X*X;
                Ysquare := Y*Y;
                Y := 2*X*Y + Q[row];
                X := Xsquare - Ysquare + P;
                inc(color);
            end;
            if color_option = 0 then
            begin
                i := 0;
                while (color > colors[i,0]) do
                    inc(i);
                if color mod 2= 0 then
                    PutPixel(col,row,colors[i][2])
                else
                    PutPixel(col,row,
                        colors[i][1]);
            end
            else
                PutPixel(col,row,color mod 16);
        end;
        P := P + deltaP;
    end;
    file1 := save_screen(0,0,639,349,file_name);
    gotoxy(1,24);
    file_name := Concat(file_string[0],file1,file2);
    write('File name is: ',file_name);
    ch1 := ReadKey;
    CloseGraph;
    until key = 3;
end.
```

The Menu Function

The Menu function first makes use of Turbo Pascal's Window procedure to set the window dimensions to that of a full screen. It then sets the background color to black and clears the screen. Depending upon the value passed in the types parameter, it then sets up a window of the proper size, fills that window with a background color of dark blue, and then writes a heading.

Next, the function enters a loop in which it remains until the Enter key is struck. It begins this loop by displaying lines of text describing several options. One of these (the first when the loop is first entered) is displayed in reverse video.

Next, a key entry is read from the keyboard. All entries except the up and down arrows and the *Enter* key are ignored. The down arrow causes the options to be redisplayed with the next one down in reverse video. When the final one is reached, the next strike of the down arrow selects the first option. Similarly, the up arrow causes cycling through the options in the opposite direction. When the *Enter* key is struck, the function is exited, returning the number of the option selected. The program then sets up another window which permits you to set the maximum iterations parameter.

Selecting Colors

The Mandelbrot set program is arranged so that you may select either of two methods for coloring the set. One is the traditional method of cycling through the available colors (16 for the EGA and VGA high-resolution modes). The program simply uses the number of iterations that are required for the set to blow up operated on mod 16 as the color for each pixel. If you make the maximum number of iterations divisible by 16, then those points that do not blow up will be colored black, as is traditional. The other option allows you to set up to 14 ranges within the range of iterations that you have chosen and assign a color to odd and another to even numbers of iterations within each range. This is similar to a method used in a public-domain program by:

Theron Wiergena
P. O. Box 595
Muskegon, MI 49443

This fractal program is available as PC-SIG disk #1076 or if you send $20.00 to the author, he will send you a disk containing the source code in Turbo Pascal. It is a very complex program; if you are interested in studying an advanced sample of fractal programming in Turbo Pascal, by all means get it.

If you select this mode of coloring, the function *set_colors* will be run. This function first sets the window to be the whole screen and fills it with red. It then creates a smaller window and fills the window with black. An overall instruction line is written, and then column headings. Next, the function begins a loop to fill the color selection array. This loop continues until you enter an upper bound that is equal to or greater than the maximum number of iterations.

First, the function enters a zero for the lower bound. It then reads the keyboard, allowing you to select the upper bound. It then displays two blocks of color, one for the odd iterations and one for the even iterations. It then loops waiting for up arrow, down arrow, or *Enter* keys. The arrow keys cycle through the available colors in opposite directions, displaying the current selection in its appropriate color block. When you are satisfied with the color of that block, hitting the *Enter* key goes on to the second color block.

When you are satisfied with that one, you go on to define the next color range. The lower limit is already set up for you as the upper limit of the previous range. When you enter the new upper limit, the next two color blocks are displayed so that you can set them. As a precaution, if you enter an upper limit that is lower than the lower limit, the program will automatically set the upper limit to the lower limit plus one. Also, if you have entered a limit higher than the maximum number of iterations, or if you have completely filled the array, the program will set the upper limit to the maximum number of iterations.

When the upper limit is equal to the maximum number of iterations, the function terminates. The function returns an integer that indicates how many entries you have put into the color array.

When the program is drawing a Mandelbrot set, the number of iterations for each pixel is compared to the array limit entries until the place where it fits is located. The program then detemines whether the number of iterations is odd or even and selects the first or second color for that entry accordingly.

Program Details

The program begins by offering you the options of drawing the Mandelbrot set, expanding part of an existing set (existing meaning that a file of display data for that set exists), completing an unfinished set, where you have saved the partially finished display to disk, or quitting. Next, you are given a chance to specify the maximum number of iterations that you want to use. If you decided to draw the Mandelbrot set, the program permits you to select the color option that you want to use. If you select cycling, the program goes ahead to draw the Mandelbrot set. If you chose to select colors, you are given that opportunity before the set is drawn. Note that for cycling you only get to cycle through the 16 default colors, but after the set is drawn, you can use the colors program given in Chapter 3 to change each of the 16 colors to any of the 64 that are available.

If you selected the option for expanding a portion of an existing Mandelbrot set, the program will ask you for a screen number between 00 and 99. Whatever number is entered becomes part of a file name of the form *mandel##.pcx*, where ## is the screen number entered. The *restore_screen* function described in Chapter 3 is then run to display this file on the screen. If the file is not found, or if it is not a proper .pcx file, the function displays a diagnostic message the program is terminated. The *move_cursor* function is then called twice. The first time, it is a type 0, which means that the cursor is drawn as the upper left corner of a rectangle.

As the cursor moves across the Mandelbrot display, the values of *XMin* and *YMax* for the current cursor position are displayed at the bottom of the display. When the appropriate point for the corner is reached, the *Ent* key is hit, whereupon the *move_cursor* function is run again as type 1, which displays the lower right corner of a rectangle. The coordinates selected for the top left corner remain displayed on the screen, and the coordinates for the current cursor position for *XMax* and *YMin* are also displayed and change as the cursor is moved about the display. The *move_cursor* function is set up so that the lower right corner can never be set above or to the left of the top left corner. When the proper position for this corner is set, hitting the *Ent* key again causes the program to begin to generate a new Mandelbrot set, with the bounds selected by the cursor.

The algorithm for generating the Mandelbrot set is just the same as that used in the program shown in Figure 12-1. However, when the program is interrupted by a keystroke, instead of exiting immediately, the program first saves the partially completed display in a disk file, using the *save_screen* function described in Chapter 3. This function always creates a new file name for the screen being saved.

The screen is then saved to the designated file, with the ending column limit variable set to the column number that was most recently completed by the Mandelbrot-generating part of the program. The .pcx file format saves the partially completed display and the color information on each of the 16 palettes, but, if we are not in the color cycling mode, more information is needed if we are to continue generating this incomplete display at some time in the future. Particularly, we need the contents of the colors array, the number of color array entries, and the maximum number of iterations. Therefore, in this color mode, these items are saved in a special palette file having the same number as the Mandelbrot file.

When you select the option to complete an unfinished display, the program gives you a chance to specify which color mode you were using. If it was color cycling, the program goes ahead and reads the display file and then continues from where it left off. If you selected the color specifying mode, the program reads the data from the palette file and then proceeds from where it left off. When a Mandelbrot set is completed, the program saves it to disk, automaticaaly adjusting the number of the disk file so as not to write over an existing display file. The program then loops back to the beginning. It will continue generating Mandelbrot sets until you select the Quit option. Expansions of the Mandelbrot set are shown in Plates 6 through 12 (in the color section). Figure 12-3 shows the bounds used for the various color pictures. Figure 12-4 shows how the color array was set for those figures when using the color cycling technique.

Figure 12-3. Parameters for Mandelbrot Set Color Plates

Plate #	XMin	XMax	YMin	YMax
4	-2.0	1.2	-1.2	1.2
A	-0.717997	-0.592801	0.395415	0.505444
5	-0.702973	-0.642879	0.374785	0.395415
6	-0.691060	-0.690906	0.387103	0.387228
7	-0.793114	-0.723005	0.037822	0.140974
B	-0.707981	-0.627856	0.477937	0.367908
C	-1.7790863	-1.778363	0.006630	0.007241
8	-0.745465	-0.745387	0.112896	0.113034
9	-0.745464	-0.745388	0.112967	0.113030

(Note: A, B, and C are not illustrated in the color section)

Figure 12-4. Color Information for Generating Color Plates

Palette #	0	1	2	3	4	5	6	7	8	9	10	11	12	13	14	15
Color Numbers																
Plate 4	0	1	2	3	4	5	20	7	56	57	58	59	60	61	62	63
A	0	1	2	3	4	5	20	7	56	57	58	59	60	61	62	63
Plate 6	0	1	2	3	4	5	20	7	56	57	58	59	60	61	62	63
Plate 7	0	1	2	3	4	5	20	7	56	57	58	59	60	61	62	63
B	9	1	2	62	2	2	2	7	56	57	16	29	56	56	56	62
C		1	32	12	4	36	37	38	7	56	57	58	59	60	61	62	63

(Note: A, B, and C are not illustrated in the color section)

Color Table for Plate 4

Lower Bound	Upper Bound	Color #1	Color #2
0	20	1	1
20	30	14	9
30	35	4	4
35	50	12	12
50	100	14	14
100	512	0	0

Color Table for A

Lower Bound	Upper Bound	Color #1	Color #2
0	20	1	2
20	30	3	11
30	35	4	12
35	50	5	15
50	100	10	14
100	512	0	0
200	256	0	0

Color Table for B

Lower Bound	Upper Bound	Color #1	Color #2
0	10	1	9
10	20	2	10
20	30	4	12
30	35	5	13
35	50	6	14
50	100	11	3
100	200	14	15
200	256	0	0

Precision Considerations

You will observe that the values of $X, delta X$, etc., used in the program of Figure 12-2, are classified as *real*. The precision of numbers available in Turbo Pascal determines how far we can expand the Mandelbrot set before the calculations begin to break down to the point that the pictures are so distorted as to be worthless. If you plan to expand the Mandelbrot set further than shown in the color plates, you can change the real type in the program to *double* or *extended* and obtain more detail at high magnifications. This may or not have a noticeable affect upon the speed with which your program generates Mandelbrot sets, depending upon whether you are using a math coprocessor and what kind of hardware system you have. You also need to be aware that *XMax*, *YMax*, *XMin*, and *YMin* are limited to the *single* type, so if you reach a magnification where you need more precision in these parameters, you are in big trouble.

Using Other Color Combinations

We've already mentioned that you can use the *colors* program to modify the colors of your Mandelbrot set displays, once they have been saved to disk files. Just in case you would like to see what these shades look like before you operate upon your Mandelbrot displays, Figure 12-5 lists a program that displays the default shades of the sixteen EGA palettes, and permits changing each one to any of the 64 available colors. The palette number is given above each color block, and as a palette is changed, the selected color number appears below the color block. When you have chosen the desired shades, you can exit this program, but first copy down the palette and color numbers for use in your program. If you want to, you can assign the same color to two or three adjacent palettes, which will reduce the amount of detail in your display, but may enhance the beauty of the picture and/or emphasize certain details that you wish to stress.

Figure 12-5. Program to Display and Change EGA Palette Colors

```
program view_palette;

uses CRT,Graph,Fractal;

const
    color: integer = 0;

var
```

```
      i,palet: integer;
begin
    clrscr;
    GotoXY(3,6);
    write('0   1   2   3   4   5   6   7   8   9   10',
        '  11  12  13  14  15');
    GotoXY(2,7);
    for i:=0 to 15 do
    begin
        TextColor(i);
        write(char($DB),char($DB),char($DB),char($DB));
    end;
    for i:= 0 to 15 do
    begin
        GotoXY(3+4*i,8);
        write(PALETTE[i]);
    end;
    while color<64 do
    begin
        GotoXY(20,22);
        write('                          ');
        GotoXY(20,23);
        write('                          ');
        GotoXY(20,22);
        write('Enter palette number: ');
        Readln(palet);
        gotoxy(20,23);
        write('Enter color number: ');
        Readln(color);
        if color < 64 then
        begin
            setEGApalette(palet,color);
            GotoXY(3+4*palet,8);
            write('   ');
            GotoXY(3+4*palet,8);
            write(color);
        end;
    end;
end.
```

Other Mandelbrot-like Sets

Although the Mandelbrot set has received all of the publicity, it is the mapping of only a single iterated function, namely that of Equation 12-1. It is not as widely known that for every iterated function, there is a set similar to the Mandelbrot set. In Chapter 14, we will be looking at dragon curves, and in Chapter 15 at phoenix curves. These curves are so-named because plotting them with the same mapping as used for Julia curves in the next chapter gives pictures that are similar to the shape of dragons and phoenixes respectively. Each of these sets has a Mandelbrot-like map. Figure 12-6 is a program to generate the Mandelbrot-like set for dragon curves. The resulting map is shown in Plate 10 (in the color section). As is the case with the Mandelbrot set, selecting interesting-looking points along the boundaries of this curve gives the values to use to generate interesting dragon curves.

Now suppose that we do exactly the same thing for the phoenix curve. (This curve will be discussed in more detail in Chapter 15.) We find that the resulting Mandelbrot- like set does not seem to be anything like a map of phoenix curves; the original phoenix curve is at coordinates that are not even near the set boundary. It turns out that we've just happened to be lucky so far. To create a set that is a map for its associated curves, we need to start the iterated equation each time with values that cause the derivative of the iterated equation to be zero. Previously, this just happened to be the case. For the set associated with the phoenix curves, we need to set the initial value of x so that:

$$x = -q/2 \qquad\qquad \text{(Equation 12-4)}$$

This makes things come out right. Figure 12-7 is a program to generate the Mandelbrot-like set for phoenix curves. This set is shown in Plate 11.

Figure 12-6. Program to Generate Mandelbrot-like Set for Dragon Curves

```
program csdragon;

uses CRT,Graph,Fractal;
```

```
const
    maxcol: integer = 639;
    maxrow: integer = 349;
    max_size: real = 4;
    file_string: array[0..1] of string[7] =
        (('drgset'),('dstpal'));
    file2: string[5] = '.pcx';
    file3: string[5] = '.pal';
    rect: array[1..4] of PointType = ((x: 0; y: 0),
                                      (x: 639; y: 0),
                                      (x: 639; y: 349),
                                      (x: 0; y: 349));

var
    graphDriver,GraphMode,i,j,key,generator_size,
        level,init_size,file_no,color,row,col,error,
        start_col,end_mask, m,color_option,max_iterations: integer;
    xtemp,ytemp,deltaP,deltaQ,X,Y,Xsquare,Ysquare,P: real;
        ch1: char;
    colors: array[0..14,0..2] of integer;
    file1: string[3];
    file_name,file_name2: string[12];
    f: file of byte;
    f_pal: file of integer;
    Q: array[0..349] of real;

function menu(types: integer): integer;
const
    TextData: array[0..3] of string[45] =
        (('Generate the Mandelbrot-like Set for Dragons'),
         ('Expand a Section of an Existing Dragon Set'),
         ('Complete an Unfinished Dragon Set'),
         ('Quit'));
    TextData2: array[0..1] of string[45] =
        (('Select colors and ranges'),
         ('Cycle through colors in order'));

var
    i,k: integer;
    ch1: char;
begin
    k := 0;
    Window(1,1,80,25);
    TextBackground(0);
    clrscr;
    if types = 0 then
    begin
        Window(10,6,59,13);
        m := 3;
    end
    else
    begin
        Window(10,6,59,11);
        m := 1;
    end;
    TextBackground(1);
```

```
        clrscr;
        GotoXY(3,2);
        TextColor(10);
        write('Make your choice with the up and down arrows:');
        repeat
            for i:=0 to m do
            begin
                if i = k then
                begin
                    TextColor(1);
                    TextBackground(15);
                end
                else
                begin
                    TextColor(15);
                    TextBackground(1);
                end;
                GotoXY(3,i+4);
                if types = 0 then
                    write(TextData[i])
                else
                    write(TextData2[i]);
            end;
            ch1:= ReadKey;
            if ch1 = char($00) then
            begin
                ch1 := ReadKey;
                case ch1 of
                    'P': begin
                            if k = 3 then
                                k := 0
                            else
                                inc(k);
                        end;
                    'H': begin
                            if k = 0 then
                                k := 3
                            else
                                dec(k);
                        end;
                end;
            end;
        until ch1 = char($0D);
        window(1,1,80,25);
        menu := k;
end;

function set_colors: integer;

const
    color_set: array[0..1] of integer = (1,1);

var
    i,k: integer;
    ch1: char;
begin
```

```
TextBackground(4);
clrscr;
    k := 0;
Window(10,4,70,22);
TextBackground(0);
clrscr;
GotoXY(3,1);
TextColor(10);
writeln('Enter Upper Limit Number and use Arrows to Set
    Colors:');
writeln;
TextColor(14);
write('  Start Iters    End Iters    Color #1  Color #2');
writeln;
i := 0;
colors[0,0] := 0;
colors[1,0] := 0;
while colors[i,0] < max_iterations do
begin
    inc(i);
    TextColor(15);
    GotoXY(6,i+5);
    write(colors[i-1,0]);
    GotoXY(20,i+5);
    Read(colors[i,0]);
    if colors[i,0] <= colors[i-1,0] then
    begin
        colors[i,0] := colors[i-1,0] + 1;
        GotoXY(20,i+5);
        write(colors[i,0]);
    end;
    if (i = 13) or (colors[i,0] > max_iterations)
    then
    begin              colors[i,0] := max_iterations;
        GotoXY(20,i+5);
        write(colors[i,0]);
    end;
    GotoXY(30,i+5);
    TextColor(color_set[0]);
    write(char($DB),char($DB),char($DB),char($DB),
        char($DB),char($DB));
    TextColor(color_set[1]);
    GotoXY(40,i+5);
    write(char($DB),char($DB),char($DB),char($DB),
        char($DB),char($DB));
    for k:= 0 to 1 do
    begin
        GotoXY(36+ 10*k,i+5);
        repeat
            ch1:= ReadKey;
            if ch1 = char($00) then
            begin
                ch1 := ReadKey;
                case ch1 of
                    'P': begin
                        if color_set[k] = 15 then
```

```
                                        color_set[k] := 0
                            else
                                inc(color_set[k]);
                            end;
                        'H': begin
                                if color_set[k] = 0 then
                                    color_set[k] := 15
                                else
                                    dec(color_set [k]);
                            end;
                    end;
                end;
                GotoXY(30+10*k,i+5);
                TextColor(color_set[k]);
                write(char($DB),char($DB),char($DB),
                    char($DB),char($DB),
                    char($DB));
            until ch1 = char($0D);
        end;
        colors[i,1] := color_set[0];
        colors[i,2] := color_set[1];
        set_colors := i;
    end;
    Window(1,1,80,25);
    clrscr;
    TextColor(15);
end;

begin
    DirectVideo := false;
    GraphDriver := 4;
    GraphMode := EGAHi;
    repeat
        key := menu(0);
        if key = 3 then
            exit;
        if key < 2 then
        begin
            TextBackground(0);
            clrscr;
            Window(15,8,66,11);
            TextBackground(4);
            clrscr;
            GotoXY(3,2);
            TextColor(15);
            write('Enter number of iterations desired
                (16-512): ');
            readln(max_iterations);
            if max_iterations < 16 then
                max_iterations := 16;
            if max_iterations > 512 then
                max_iterations := 512;
        end;
        color_option := menu(1);
        if (key < 2) and (color_option = 0) then
            end_mask :=  set_colors;
```

```
TextBackground(0);
clrscr;
Window(15,8,66,11);
TextBackground(2);
clrscr;
GotoXY(3,2);
TextColor(15);
if key <> 0 then
begin
    write('Enter "DRGSET" file number (00 - 99): ');
    readln(file_no);
    Str(file_no,file1);
    Window(1,1,80,25);
    if file_no < 10 then
        file_name := Concat
            (file_string[0],'0',file1,file2)
    else
        file_name := Concat
            (file_string[0],file1,file2);
    error := restore_screen(file_name);
    if error = 0 then
        exit
end;
CURSOR_X := 0;
CURSOR_Y := 0;
case key of
0: begin
    start_col := 0;
    XMax := 4.2;
    XMin := -2.2;
    YMax := 1.5;
    YMin := -1.5;
    file_name := Concat(file_string[0],'00',file2);
    InitGraph(GraphDriver,GraphMode,'');
    SetFillStyle(1,7);
    FillPoly(4,rect);
    end;
2: begin
    if (error = 639) or (error = 0) then
        exit;
    start_col := 8 * (error div 8);
    Assign(f,file_name);
    Erase(f);
    if color_option = 0 then
    begin
    if file_no < 10 then
        file_name := Concat
            (file_string[1],'0',file1,file3)
    else
        file_name := Concat
            (file_string[1],file1,file3);
        Assign(f_pal,file_name2);
        Reset(f_pal);
        for i:= 0 to 14 do
            for j:= 0 to 2 do
                read(f_pal,colors[i,j]);
```

279

```
          read(f_pal,max_iterations);
          read(f_pal,end_mask);
          Close(f_pal);
      end;
    end;
    1: begin
          move_cursor(0,15,0,0);
          move_cursor(1,15,CURSOR_X,CURSOR_Y);
          XMax := TXMax;
          XMin := TXMin;
          YMax := TYMax;
          YMin := TYMin;
          start_col := 0;
          SetFillStyle(1,7);
          FillPoly(4,rect);
        end;
    end;
deltaP := (XMax - XMin)/(maxcol);
deltaQ := (YMax - YMin)/(maxrow);
Q[0] := YMax;
for row:=1 to maxrow do
    Q[row] := Q[row-1] - deltaQ;
    P := XMin + start_col * deltaP;
for col:=start_col to maxcol do
begin
    if KeyPressed then
    begin
        file1 :=
            save_screen(0,0,col,349,file_name);
        gotoxy(1,24);
        file_name := Concat
            (file_string[0],file1,file2);
        write('File name is: ',file_name);
        if color_option = 0 then
        begin
            file_name := Concat
                (file_string[1],file1,file3);
            Assign(f_pal,file_name);
            Rewrite(f_pal);
            for i:= 0 to 14 do
                for j:= 0 to 2 do
                    write(f_pal,colors[i,j]);
            write(f_pal,max_iterations);
            write(f_pal,end_mask);
            Close(f_pal);
        end;
        exit;
    end;
    for row:=0 to maxrow do
    begin
        X := 0.50;
        Y := 0.0;
        color := 0;
        Xsquare := 0;
        Ysquare := 0;
        color := 1;
```

```
            while (color<max_iterations) and
                ((X*X + Y*Y) < max_size) do
            begin
                Xtemp := (Y - X)*(Y + X) + X;
                Ytemp := X * Y;
                Ytemp := Ytemp + Ytemp - Y;
                X := P * Xtemp + Q[row] * Ytemp;
                Y := Q[row] * Xtemp - P * Ytemp;
                inc(color);
            end;
            if color_option = 0 then
            begin
                i := 0;
                while (color > colors[i,0]) do
                    inc(i);
                if color mod 2= 0 then
                    PutPixel(col,row,colors[i][2])
                else
                    PutPixel(col,row,colors[i][1]);
            end
            else
                PutPixel(col,row,color mod 16);
        end;
        P := P + deltaP;
    end;
    file1 := save_screen(0,0,639,349,file_name);
    gotoxy(1,24);
    file_name := Concat(file_string[0],file1,file2);
    write('File name is: ',file_name);
    ch1 := ReadKey;
    CloseGraph;
  until key = 3;
end.
```

Figure 12-7. Program to Generate Mandelbrot-like Set
for Phoenix Curves

```
program csphenix;

uses CRT,Graph,Fractal;

const
    maxcol: integer = 639;
    maxrow: integer = 349;
    max_size: real = 4;
    file_string: array[0..1] of string[7] =
        (('pheset'),('phepal'));
    file2: string[5] = '.pcx';
    file3: string[5] = '.pal';
    rect: array[1..4] of PointType = ((x: 0; y: 0),
                                      (x: 639; y: 0),
                                      (x: 639; y: 349),
                                      (x: 0; y: 349));
```

```
var
    graphDriver,GraphMode,i,j,key,generator_size,level,
        init_size, file_no,color,row,col,error,
        start_col,end_mask,m,color_option,max_iterations:
        integer;
    deltaP,deltaQ,X,Xi,Y,Yi,Xtemp,Xitemp,Xsquare,
        Xisquare,Ysquare,P: real;
    ch1: char;
    colors: array[0..14,0..2] of integer;
    file1: string[3];
    file_name,file_name2: string[12];
    f: file of byte;
    f_pal: file of integer;
    Q: array[0..349] of real;


function menu(types: integer): integer;

const
    TextData: array[0..3] of string[45] =
        (('Generate the Mandelbrot-like Set for Phoenix
            Curves'),
        ('Expand a Section of an Existing Phoenix '),
        ('Complete an Unfinished Phoenix Set'),
        ('Quit'));
    TextData2: array[0..1] of string[45] =
        (('Select colors and ranges'),
        ('Cycle through colors in order'));

var
    i,k: integer;
    ch1: char; begin
    k := 0;
    Window(1,1,80,25);
    TextBackground(0);
    clrscr;
    if types = 0 then
    begin
        Window(10,6,59,13);
        m := 3;
    end
    else
    begin
        Window(10,6,59,11);
        m := 1;
    end;
    TextBackground(1);
    clrscr;
    GotoXY(3,2);
    TextColor(10);
    write('Make your choice with the up and down arrows:');
    repeat
        for i:=0 to m do
        begin
            if i = k then
            begin
```

```pascal
                TextColor(1);
                TextBackground(15);
            end
            else
            begin
                TextColor(15);
                TextBackground(1);
            end;
            GotoXY(3,i+4);
            if types = 0 then
                write(TextData[i])
            else
                write(TextData2[i]);
        end;
        ch1:= ReadKey;
        if ch1 = char($00) then
        begin
            ch1 := ReadKey;
            case ch1 of
                'P': begin
                        if k = 3 then
                            k := 0
                        else
                            inc(k);
                    end;
                'H': begin
                        if k = 0 then
                            k := 3
                        else
                            dec(k);
                    end;
            end;
        end;
    until ch1 = char($0D);
    window(1,1,80,25);
    menu := k;
end;

function set_colors: integer;

const
    color_set: array[0..1] of integer = (1,1);

var
    i,k: integer;
    ch1: char;

begin
    TextBackground(4);
    clrscr;
        k := 0;
    Window(10,4,70,22);
    TextBackground(0);
    clrscr;
    GotoXY(3,1);
    TextColor(10);
```

```
writeln('Enter Upper Limit Number and use Arrows to Set
    Colors:');
writeln;
TextColor(14);
write('  Start Iters   End Iters   Color #1  Color #2');
writeln;
i := 0;
colors[0,0] := 0;
colors[1,0] := 0;
while colors[i,0] < max_iterations do
begin
    inc(i);
    TextColor(15);
    GotoXY(6,i+5);
    write(colors[i-1,0]);
    GotoXY(20,i+5);
    Read(colors[i,0]);
    if colors[i,0] <= colors[i-1,0] then
    begin
        colors[i,0] := colors[i-1,0] + 1;
        GotoXY(20,i+5);
        write(colors[i,0]);
    end;
    if (i = 13) or (colors[i,0] > max_iterations)
    then
    begin
        colors[i,0] := max_iterations;
        GotoXY(20,i+5);
        write(colors[i,0]);
    end;
    GotoXY(30,i+5);
    TextColor(color_set[0]);
    write(char($DB),char($DB),char($DB),char($DB),
        char($DB),char($DB));
            TextColor(color_set[1]);
    GotoXY(40,i+5);
    write(char($DB),char($DB),char($DB),char($DB),
        char($DB),char($DB));
    for k:= 0 to 1 do
    begin
        GotoXY(36+ 10*k,i+5);
        repeat
            ch1:= ReadKey;
            if ch1 = char($00) then
            begin
                ch1 := ReadKey;
                case ch1 of
                    'P': begin
                        if color_set[k] = 15 then
                            color_set[k] := 0
                        else
                            inc(color_set[k]);
                      end;
                    'H': begin
                        if color_set[k] = 0 then
                            color_set[k] := 15
```

```
                                        else
                                            dec(color_set [k]);
                                    end;
                            end;
                    end;
                    GotoXY(30+10*k,i+5);
                    TextColor(color_set[k]);
                    write(char($DB),char($DB),char($DB),
                        char($DB),char($DB),char($DB));
                until ch1 = char($0D);
            end;
            colors[i,1] := color_set[0];
            colors[i,2] := color_set[1];
            set_colors := i;
        end;
    Window(1,1,80,25);
    clrscr;
    TextColor(15);
end;

begin
    DirectVideo := false;
    GraphDriver := 4;
    GraphMode := EGAHi;
    repeat
        key := menu(0);
        if key = 3 then
            exit;
        if key < 2 then
        begin
            TextBackground(0);
            clrscr;
            Window(15,8,66,11);
            TextBackground(4);
            clrscr;
            GotoXY(3,2);
            TextColor(15);
            write('Enter number of iterations desired
                (16-512): ');
            readln(max_iterations);
            if max_iterations < 16 then
                max_iterations := 16;
            if max_iterations > 512 then
                max_iterations := 512;
        end;
        color_option := menu(1);
        if (key < 2) and (color_option = 0) then
            end_mask :=  set_colors;
        TextBackground(0);
        clrscr;
        Window(15,8,66,11);
        TextBackground(2);
        clrscr;
        GotoXY(3,2);
        TextColor(15);
        if key <> 0 then
```

```
begin
    write('Enter "PHESET" file number (00 - 99): ');
    readln(file_no);
    Str(file_no,file1);
    Window(1,1,80,25);
    if file_no < 10 then
        file_name := Concat
            (file_string[0],'0',file1,file2)
    else
        file_name := Concat
            (file_string[0],file1,file2);
    error := restore_screen(file_name);
    if error = 0 then
        exit
end;
CURSOR_X := 0;
CURSOR_Y := 0;
case key of
0: begin
    start_col := 0;
    XMax := 1.5;
    XMin := -2.1;
    YMax := 2.0;
    YMin := -2.0;
    file_name := Concat
        (file_string[0],'00',file2);
    InitGraph(GraphDriver,GraphMode,'');
    SetFillStyle(1,7);
    FillPoly(4,rect);
    end;
2: begin

    if (error = 639) or (error = 0) then
        exit;
    start_col := 8 * (error div 8);
    Assign(f,file_name);
    Erase(f);
    if color_option = 0 then
    begin
    if file_no < 10 then
        file_name := Concat
            (file_string[1],'0',file1,file3)
    else
        file_name := Concat
            (file_string[1],file1,file3);
        Assign(f_pal,file_name2);
        Reset(f_pal);
        for i:= 0 to 14 do
            for j:= 0 to 2 do
                read(f_pal,colors[i,j]);
        read(f_pal,max_iterations);
        read(f_pal,end_mask);
        Close(f_pal);
    end;
    end;
    1: begin
```

```
                move_cursor(0,15,0,0);
                move_cursor(1,15,CURSOR_X,CURSOR_Y);
                XMax := TXMax;
                XMin := TXMin;
                YMax := TYMax;
                YMin := TYMin;
                start_col := 0;
                SetFillStyle(1,7);
                FillPoly(4,rect);
            end;
        end;
    deltaP := (XMax - XMin)/(maxcol);
    deltaQ := (YMax - YMin)/(maxrow);
    Q[0] := YMax;
    for row:=1 to maxrow do
        Q[row] := Q[row-1] - deltaQ;
        P := XMin + start_col * deltaP;
    for col:=start_col to maxcol do
    begin
        if KeyPressed then
        begin
            file1 :=
                save_screen(0,0,col,349,file_name);
            gotoxy(1,24);
            file_name := Concat
                (file_string[0],file1,file2);
            write('File name is: ',file_name);
            if color_option = 0 then
            begin
                file_name := Concat
                    (file_string[1],file1,file3);
                Assign(f_pal,file_name);
                Rewrite(f_pal);
                for i:= 0 to 14 do
                    for j:= 0 to 2 do
                        write(f_pal,colors[i,j]);
                write(f_pal,max_iterations);
                write(f_pal,end_mask);
                Close(f_pal);
            end;
            exit;
        end;
        for row:=0 to maxrow do
        begin
            X := -0.5*Q[row];
            Xi := 0;
            Y := 0;
            Yi := 0;
            color := 0;
            Xsquare := 0;
            Xisquare := 0;
            while (color<max_iterations) and
                (Xsquare + Xisquare < max_size) do
            begin
                Xsquare := X*X;
                Xisquare := Xi*Xi;
```

```
                    Xtemp := Xsquare - Xisquare + P
                        + Q[row]*Y;
                    Xitemp := 2*X*Xi + Q[row]*Yi;
                    Y := X;
                    Yi := Xi;
                    X := Xtemp;
                    Xi := Xitemp;
                    inc(color);
                end;
                if color_option = 0 then
                begin
                    i := 0;
                    while (color > colors[i,0]) do
                        inc(i);
                    if color mod 2= 0 then
                        PutPixel(col,row,colors[i][2])
                    else
                            PutPixel(col,row,colors[i][1]);
                end
                else
                        PutPixel(col,row,color mod 16);
            end;
            P := P + deltaP;
        end;
        file1 := save_screen(0,0,639,349,file_name);
        gotoxy(1,24);
        file_name := Concat(file_string[0],file1,file2);
        write('File name is: ',file_name);
        ch1 := ReadKey;
        CloseGraph;
    until key = 3;
end.
```

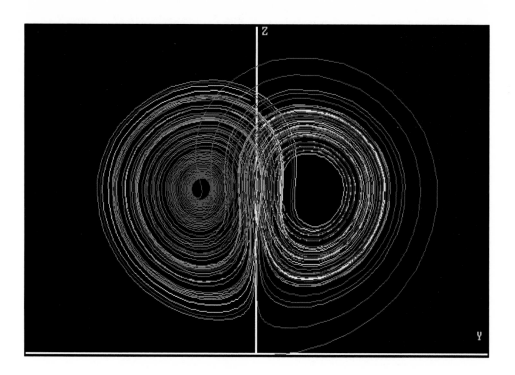

Plate 1: Lorenz Attractor Projected on YZ Plane

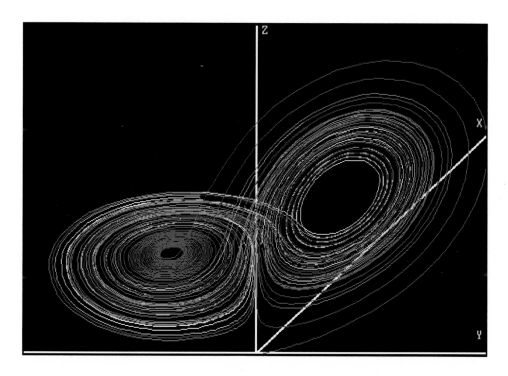

Plate 2: Three-Dimensional View of Lorenz Attractor

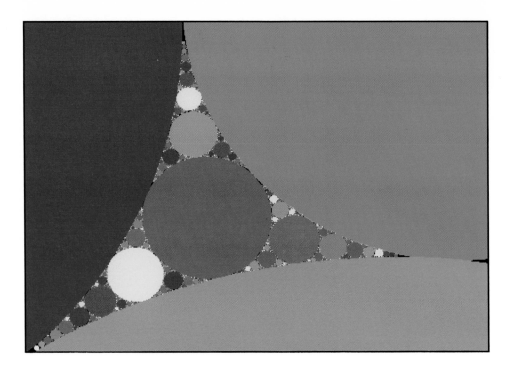

Plate 3: Apollonian Packing of Circles

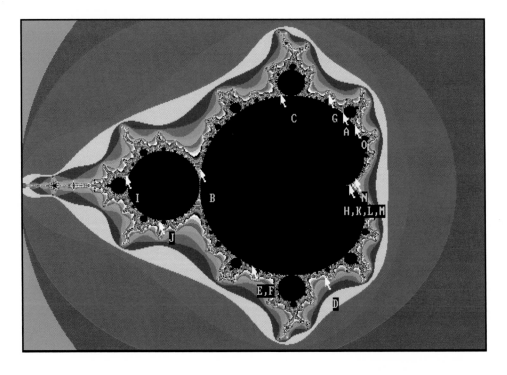

Plate 4: Mandelbrot Set Showing Locations for Julia Sets

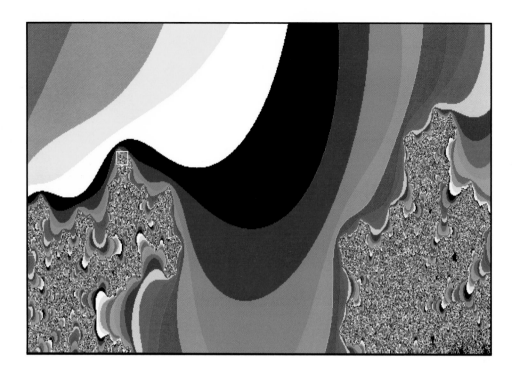

Plate 5: Expansion of an Area in Plate 4

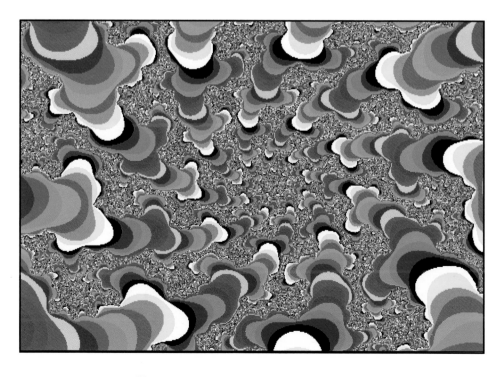

Plate 6: Expansion of an Area in Plate 5

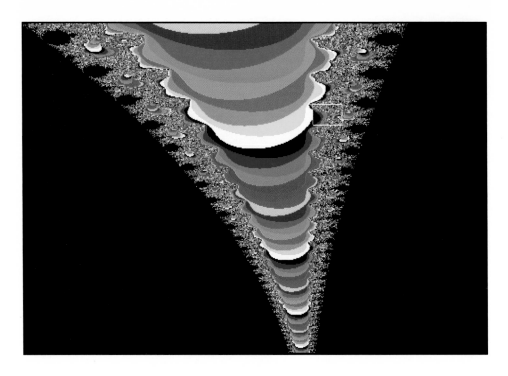

Plate 7: Expansion of an Area in Plate 4

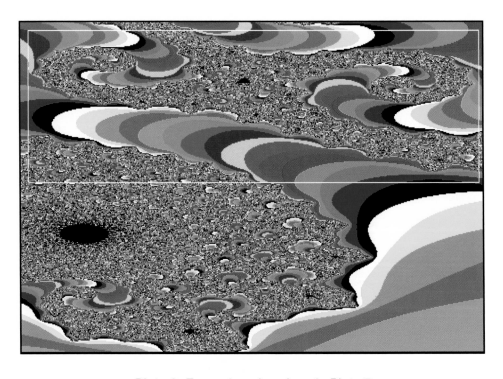

Plate 8: Expansion of an Area in Plate 7

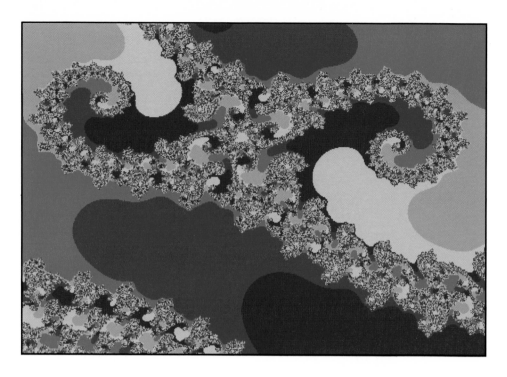

Plate 9: Expansion of Marked Area in Plate 8 with Color Changes

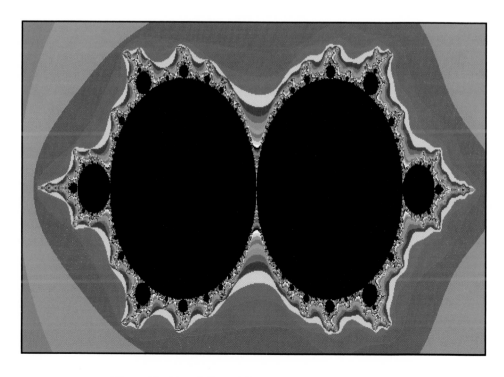

Plate 10: Mandelbrot-Like Set for Dragon Curves

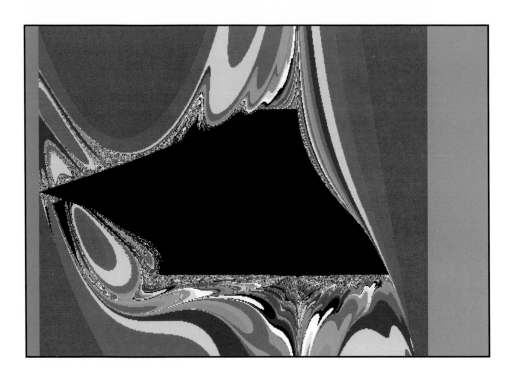

Plate 11: Mandelbrot-Like Set for Phoenix Curves

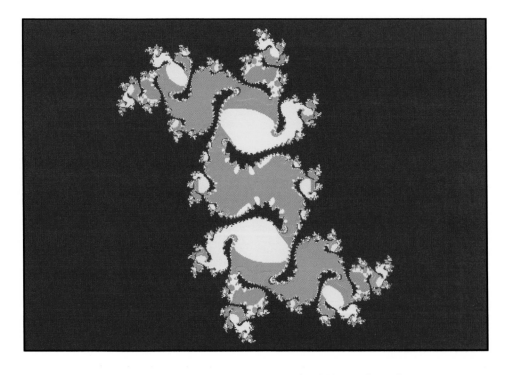

Plate 12: Julia Set from Location A of Mandelbrot Set

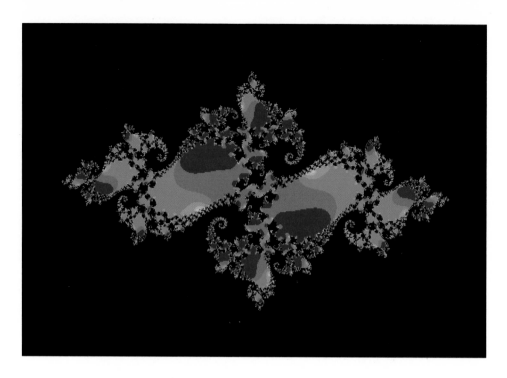

Plate 13: Julia Set from Location B of Mandelbrot Set

Plate 14: Julia Set from Location D of Mandelbrot Set

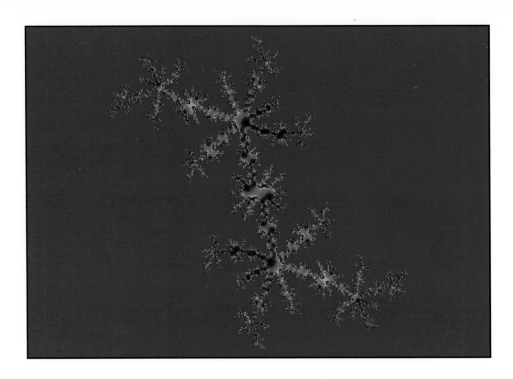

Plate 15: Julia Set from Location G of Mandelbrot Set

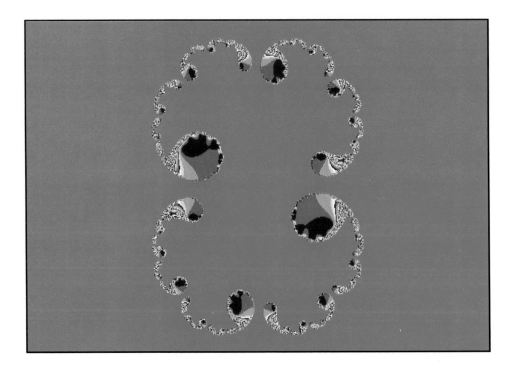

Plate 16: Julia Set from Location H of Mandelbrot Set

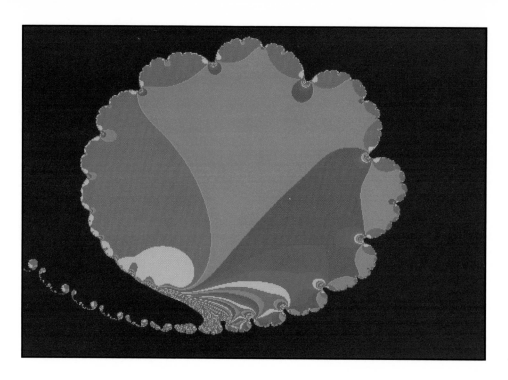

Plate 17: Julia Set from Location L of Mandelbrot Set

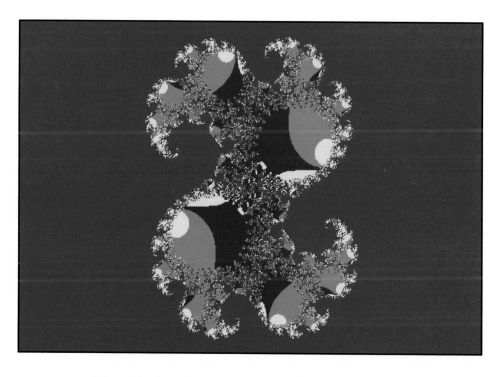

Plate 18: Julia Set from Location N of Mandelbrot Set

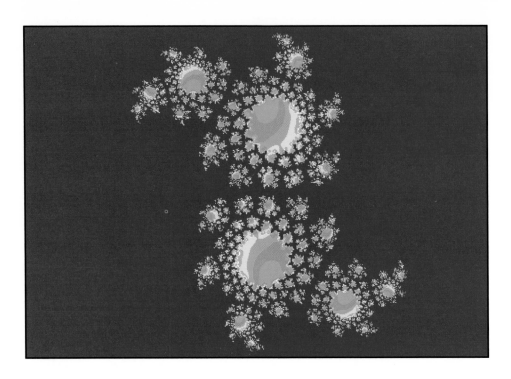

Plate 19: Julia Set from Location O of Mandelbrot Set

Plate 20: Twin Dragon Curve

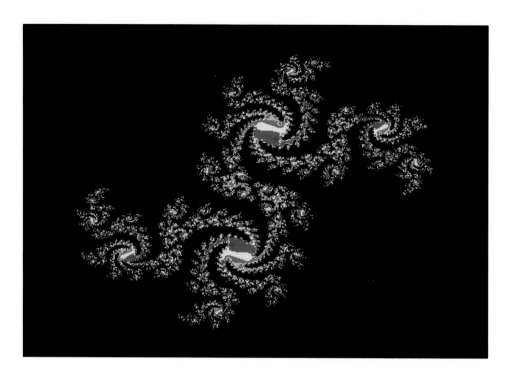

Plate 21: Self-Squared Dragon

Plate 22: San Marcos Dragon

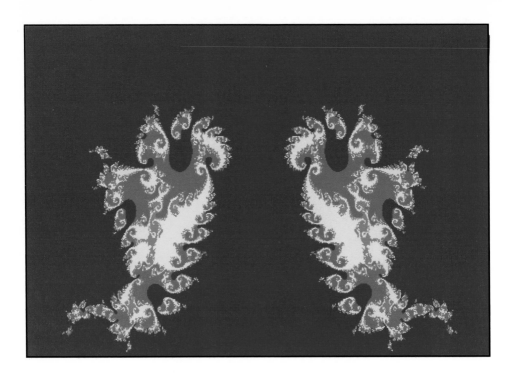

Plate 23: Original Phoenix Curve

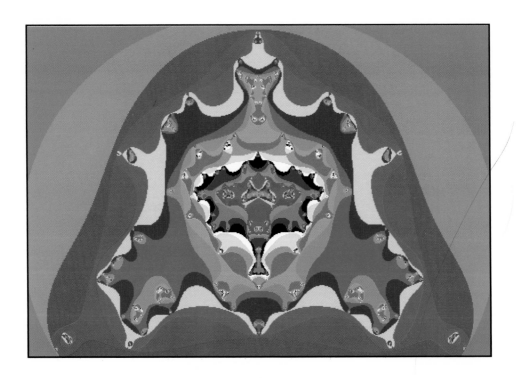

Plate 24: Phoenix Curve

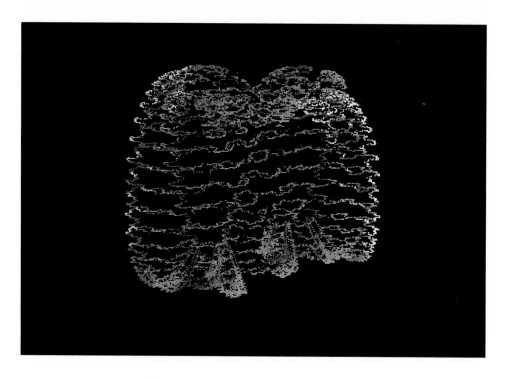

Plate 25: Three-Dimensional Dragon

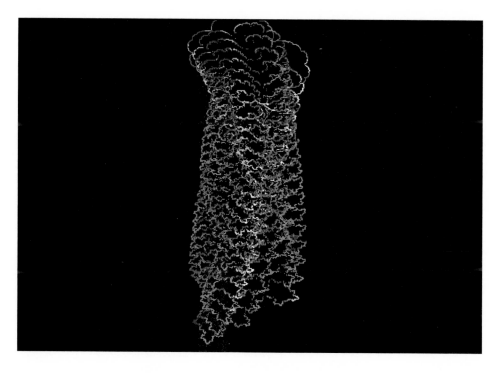

Plate 26: Three-Dimensional Dragon

Plate 27: Solution of $z^3 - 1 = 0$ by Newton's Method

Plate 28: Solution of $z^3 - 2z - 5 = 0$ by Newton's Method

Plate 29: Oak Creek Canyon

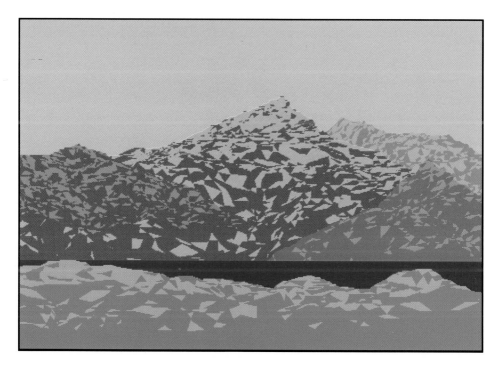

Plate 30: Pike's Peak at Sunrise

Plate 31: Earth Viewed from the Moon

Plate 32: Three-Dimensional Ferns

13

Julia Sets

Chapter 12 discussed the Mandelbrot set, which is produced by plotting the values of the iterated equation:

$$z_{n+1} = z_n^2 + c \qquad \text{(Equation 13-1)}$$

with the x display coordinate corresponding to the real part of c over a selected range, and the y display coordinate corresponding to the imaginary part of c over a selected range. For each calculation, the beginning value of z is taken as zero. It was pointed out that there is another way in which this equation can be mapped onto the display, which produces the Julia sets. For this other technique, a particular value is selected for c. The equation is then processed for various values of z_0 over a selected range. The x coordinate of the display corresponds to the real part of z_0 and the y coordinate corresponds to the imaginary part of z_0. It was also mentioned that the Mandelbrot set forms a sort of map of all of the possible Julia sets.

Once you have generated and saved the Mandelbrot set shown in Plate 4, you can use the Julia set program and move the cursor around the Mandelbrot set to find locations which will generate interesting Julia sets. The coordinates of a number of these are shown later in Figure 13-2. Seven of the more interesting Julia sets are displayed in color in Plates 12 through 19 (in the color section). The first thing to note is where interesting Julia sets occur on the Mandelbrot map. The most interesting patterns occur for values of c that are very close to the boundary of the Mandelbrot set, and usually also near a cusp. It is also interesting, but not very useful, to note that if a particular portion of the Mandelbrot set is expanded to a large enough scale, the resulting pattern is very much like the Julia set for the value of c at the center of the expansion. You can verify this by selecting a value for drawing a particular Julia set, and then expanding the Mandelbrot set around the same point, using the program in Chapter 12.

Drawing Julia Sets

Figure 13-1 is a program for plotting Julia sets. The options available in this program are much like those given in the expanded Mandelbrot set program described in the previous chapter. Before beginning this program, you must have generated the basic Mandelbrot set with the Mandelbrot set program. The Julia program then gives you the options of selecting a point on the Mandelbrot set from which to draw a Julia set, expanding a Julia set, finishing an incomplete Julia set drawing, or quitting. If you want to expand a Julia set, you must first draw the fundamental Julia set by setting a point on the Mandelbrot set, and then select an area on your new Julia set for expansion.

Next, the program asks you to enter the maximum number of iterations that will be performed by the iteration loop. You then can select colors or authorize the color cycling option. The cycling option simply cycles through the 16 default palette colors; you can later use the *colors* program to change each of these palettes to any of the 64 shades of color available in the EGA/VGA high-resolution mode. If you decide to create a new Julia set, you are asked to enter two digits to form a file name. The program then displays the Mandelbrot set having the file name *mandel##.pcx*, where ## are the two digits that you have just entered. If such a file doesn't exist, or is not a legitimate .pcx file, the program prints a diagnostic message and terminates. If a Mandelbrot set display does appear, you are given the opportunity to move a cursor arrow to select the P (real part of c) and Q (imaginary part of c) values for a Julia set.

As the cursor arrow moves about the screen, the values of P and Q are displayed at the bottom of the screen. The cursor moves one pixel for each time one of the arrow keys is hit. By holding down the Shift key, this movement can be increased to ten pixels per arrow key entry. If you chose to complete or expand a Julia set, you are asked to enter two numbers for the Julia set screen number. The program then attempts to display a file having the file name *julia0##.pcx*, where ## are the two numbers that you just entered. Again, if the file does not exist, or is not a legitimate *.pcx* file, the program terminates. Otherwise, the file is displayed. If the file is an incomplete display and the color selection mode was in operation, a second file is read containing the color information. The program then picks up where it left off and continues work on completing the display. If you selected the expansion mode, the program displays an upper left corner cursor,and permits you to select *XMin* and *YMax* values for your expanded display with the cursor arrows. When the cursor is correctly positioned, hitting *Ent* displays the lower right corner cursor and allows you to select the desired values of *XMax* and *YMin*.

The program then begins to run a set of nested loops very similar to those described for the Mandelbrot set program, except that P and Q are held constant and the initial values of X and Y are varied as the program generates a result for each pixel. After either the equation has blown up by exceeding the magnitude of two, or after the specified maximum number of iterations has occurred, a color is plotted to the appropriate pixel using whichever color method was selected.

Figure 13-1. Program to Generate Julia Sets

```
program julia;

uses CRT,Graph,Fractal;

const
    maxcol: integer = 639;
```

```
    maxrow: integer = 349;
    max_size: real = 4;
    file_string: array[0..3] of string[7] =
        (('mandel'),('julia0'), ('julia0'),('palett'));
    file2: string[5] = '.pcx';
    file3: string[5] = '.pal';
    file_name2: string[20] = 'julia000.pcx';
    rect: array[1..4] of PointType = ((x: 0; y: 0),
                                      (x: 639; y: 0),
                                      (x: 639; y: 349),
                                      (x: 0; y: 349));

var
    graphDriver,GraphMode,i,j,key,generator_size,
        level,init_size, file_no,color,row,col,error,
        start_col,end_mask, m,color_option,max_iterations:
        integer;
    deltaX,deltaY,X,Y,Xsquare,Ysquare,P,Q: real;
    ch1: char;
    colors: array[0..14,0..2] of integer;
    file1: string[3];
    file_name: string[12];
    f: file of byte;
    f_pal: file of integer;

function menu(types: integer): integer;

const
    TextData: array[0..3] of string[45] =
        (('Select Julia Set location from ',
        Mandelbrot Set'),
        ('Expand a Section of an Existing Julia Set'),
        ('Complete an Unfinished Julia Set'), ('Quit'));
    TextData2: array[0..1] of string[45] =
        (('Select colors and ranges'),
        ('Cycle through colors in order'));

var
    i,k: integer;
    ch1: char;
begin
    k := 0;
    Window(1,1,80,25);
    TextBackground(0);
    clrscr;
    if types = 0 then
    begin
        Window(10,6,59,13);
        m := 3;
    end
    else
```

```
begin
    Window(10,6,59,11);
    m := 1;
end;
TextBackground(1);
clrscr;
GotoXY(3,2);
TextColor(10);
write('Make your choice with the up and down arrows:');
repeat
    for i:=0 to m do
    begin
        if i = k then
        begin
            TextColor(1);
            TextBackground(15);
        end
        else
        begin
            TextColor(15);
            TextBackground(1);
        end;
        GotoXY(3,i+4);
        if types = 0 then
            write(TextData[i])
        else
            write(TextData2[i]);
    end;
    ch1:= ReadKey;
    if ch1 = char($00) then
    begin
        ch1 := ReadKey;
        case ch1 of
            'P': begin
                    if k = 3 then
                        k := 0
                    else
                        inc(k);
                end;
            'H': begin
                    if k = 0 then
                        k := 3
                    else
                        dec(k);
                end;
        end;
    end;
until ch1 = char($0D);
window(1,1,80,25);
menu := k;
end;
```

```pascal
function set_colors: integer;

const
    color_set: array[0..1] of integer = (1,1);

var
    i,k: integer;
    ch1: char;
begin
    TextBackground(4);
    clrscr;
        k := 0;
    Window(10,4,70,22);
    TextBackground(0);
    clrscr;
    GotoXY(3,1);
    TextColor(10);
    writeln('Enter Upper Limit Number and use Arrows to',
        'Set Colors:');
    writeln;
    TextColor(14);
    write('  Start Iters    End Iters    Color #1',
        '  Color #2');
     writeln;
    i := 0;
    colors[0,0] := 0;
    colors[1,0] := 0;
    while colors[i,0] < max_iterations do
    begin
        inc(i);
        TextColor(15);
        GotoXY(6,i+5);
        write(colors[i-1,0]);
        GotoXY(20,i+5);
        Read(colors[i,0]);
        if colors[i,0] <= colors[i-1,0] then
        begin
            colors[i,0] := colors[i-1,0] + 1;
            GotoXY(20,i+5);
            write(colors[i,0]);
        end;
        if (i = 13) or (colors[i,0] > max_iterations) then
        begin
            colors[i,0] := max_iterations;
            GotoXY(20,i+5);
            write(colors[i,0]);
        end;
        GotoXY(30,i+5);
        TextColor(color_set[0]);
        write(char($DB),char($DB),char($DB),char($DB),
```

```
                char($DB),char($DB));
            TextColor(color_set[1]);
            GotoXY(40,i+5);
            write(char($DB),char($DB),char($DB),char($DB),
                char($DB),char($DB));
            for k:= 0 to 1 do
            begin
                GotoXY(36+ 10*k,i+5);
                repeat
                    ch1:= ReadKey;
                    if ch1 = char($00) then
                    begin
                        ch1 := ReadKey;
                        case ch1 of
                            'P': begin
                                if color_set[k] = 15 then
                                    color_set[k] := 0
                                else
                                    inc(color_set[k]);
                                end;
                            'H': begin
                                if color_set[k] = 0
                                    then
                                    color_set[k] :=
                                        15
                                else
                                    dec(color_set
                                        [k]);
                                end;
                        end;
                    end;
                    GotoXY(30+10*k,i+5);
                    TextColor(color_set[k]);
                    write(char($DB),char($DB),char($DB),
                        char($DB), char($DB), char($DB));
                until ch1 = char($0D);
            end;
            colors[i,1] := color_set[0];
            colors[i,2] := color_set[1];
            set_colors := i;
        end;    Window(1,1,80,25);
        clrscr;
        TextColor(15);
    end;

begin
    repeat
        key := menu(0);
        if key = 3 then
            exit;
        if key < 2 then
```

```
begin
    TextBackground(0);
    clrscr;
    Window(15,8,66,11);
    TextBackground(4);
    clrscr;
    GotoXY(3,2);
    TextColor(15);
    write('Enter number of iterations desired',
        '(16-512): ');
    readln(max_iterations);
    if max_iterations < 16 then
        max_iterations := 16;
    if max_iterations > 512 then
        max_iterations := 512;
end;
color_option := menu(1);
if (key < 2) and (color_option = 0) then
    end_mask :=  set_colors;
TextBackground(0);
clrscr;
Window(15,8,66,11);
TextBackground(2);
clrscr;
GotoXY(3,2);
TextColor(15);
if key = 0 then
    write('Enter Mandelbrot set number',
        '(00 - 99): ')
else
    write('Enter Julia set number (00 - 99); ');
readln(file_no);
Str(file_no,file1);
Window(1,1,80,25);
if file_no < 10 then
    file_name := Concat
        (file_string[key],'0',file1,file2)
else
    file_name := Concat
        (file_string[key],file1,file2);
error := restore_screen(file_name);
if error = 0 then
    exit
else
begin
    CURSOR_X := 0;
    CURSOR_Y := 0;
    case key of
    0: begin
        start_col := 0;
        move_cursor(2,15,0,0);
```

```
      XMax := 1.8;
      XMin := -1.8;
      YMax := 1.2;
      YMin := -1.2;
      P := Pval;
      Q := Qval;
      SetFillStyle(1,7);
      FillPoly(4,rect);
    end;
2: begin
    if (error = 639) or (error = 0) then
        exit;
    start_col := 8 * (error div 8);
    file_name2 := Copy
        (file_name,0,length(file_name));
    Assign(f,file_name);
    Erase(f);
    P := Pval;
    Q := Qval;
    if color_option = 0 then
    begin
        if file_no < 10 then
            file_name2 := Concat
                (file_string[3],'0', file1,file3)
        else
            file_name := Concat
                (file_string[3], file1,file3);
        Assign(f_pal,file_name);
        Reset(f_pal);
        for i:= 0 to 14 do
            for j:= 0 to 2 do
                read(f_pal,colors[i,j]);
        read(f_pal,max_iterations);
        read(f_pal,end_mask);
        Close(f_pal);
    end;
    end;
1: begin
    move_cursor(0,15,0,0);
    move_cursor(1,15,CURSOR_X,CURSOR_Y);
    XMax := TXMax;
    XMin := TXMin;
    YMax := TYMax;
    YMin := TYMin;
    P := Pval;
    Q := Qval;
    start_col := 0;
    SetFillStyle(1,7);
    FillPoly(4,rect);
    end;
end;
```

```
deltaX := (XMax - XMin)/(maxcol);
deltaY := (YMax - YMin)/(maxrow);
for col:=start_col to maxcol do
begin
    if KeyPressed then
    begin
        Pval := P;
        Qval := Q;
        file1 := save_screen(0,0, col,349, file_name2);
        gotoxy(1,24);
        file_name := Concat
            (file_string[2],file1,file2);
        write('File name is: ',file_name);
        if color_option = 0 then
        begin
            file_name := Concat
                (file_string[3],file1,file3);
            Assign(f_pal,file_name);
            Rewrite(f_pal);
            for i:= 0 to 14 do
                for j:= 0 to 2 do
                    write(f_pal,colors[i,j]);
            write(f_pal,max_iterations);
            write(f_pal,end_mask);
            Close(f_pal);
        end;
        exit;
    end;
    for row:=0 to maxrow do
    begin
        X := XMin + col * deltaX;
        Y := YMax - row * deltaY;
        Xsquare := 0;
        Ysquare := 0;
        color := 0;
        while (color<max_iterations) and
            ((Xsquare + Ysquare) < max_size) do
        begin
            Xsquare := X*X;
            Ysquare := Y*Y;
            Y := 2*X*Y + Q;
            X := Xsquare - Ysquare + P;
            inc(color);
        end;
        if color_option = 0 then
        begin
            i:=0;
            while (color > colors[i,0]) do
                inc(i);
            if color mod 2= 0 then
                PutPixel(col,row,colors[i][2])
```

```
                    else
                        PutPixel(col,row,colors [i][1]);
                end
                else
                begin
                    if color >= max_iterations then
                        color := (Round((Xsquare +
                            Ysquare)*14) mod
                            14) + 1
                    else
                        color := 0;
                    PutPixel(col,row,color);
                end;
            end;
        end;
        end;
        Pval := P;
        Qval := Q;
        file1 := save_screen(0,0,639,349,file_name2);
        gotoxy(1,24);
        file_name := ConCat(file_string[2],file1,file2);
        write('File name is: ',file_name);
        ch1 := ReadKey;
        CloseGraph;
    until key = 3;
end.
```

Number of Iterations

As this program is used at points very near the border of the Mandelbrot set, a situation occurs where the precision of the computer is inadequate to prevent the value of the iterated equation from drifting off and finally blowing up if enough iterations take place, even though it would not blow up if the computer were absolutely accurate. Thus, if you enter a very large number of iterations, the program will not only be inordinately slow, but the result will be a completely blank screen, painted in the background color.

As you begin to reduce the number of iterations, you will start getting a display that has a great amount of detail, but also a large number of lost points. The number of iterations required to present a display which is a good compromise between adequate detail and reasonably complete representation of the pattern varies depending upon the exact values which you have selected for P and Q. That is why the program has been set up to allow you to enter the number of iterations as an input. A good place to begin is with 64 iterations. Then, if the display is sparse, with a large number of isolated points, you should decrease the number of iterations; whereas if the display has large blobs of the same color and appears to lack detail, you should increase the number of iterations.

Julia Set Displays

Figure 13-2 shows the parameters used in the Julia set displays, which comprise Plates 12 through 19 (see color section) and also parameters for some other Julia sets that you might want to generate on your own computer. These pictures provide a good representation of the types of Julia displays which can be generated from the various locations in the Mandelbrot set identified in Plate 4. Note that the color information tells you how to set the colors on an already generated display using the colors program. All of the displays except that in A used the color cycling mode. The information used in selecting colors for A is given in Figure 13-3.

A few comments need to be made on these displays. F and G are pictures generated using the same values of P and Q, but with different color schemes and a different number of iterations. They give a clear comparison of how detail increases, but parts of the display disappear when the number of iterations increases. J is an expansion of I, using the cursor to select a new area for enlargement. The new limits are: $XMin = -0.673239$, $XMax = -0.171831$, $YMin = 0.171928$, and $YMax = 0.402292$. M and N also have identical values of P and Q. The picture in N uses only 24 iterations. This is obviously too few in this case; too much detail is lost. The picture does provide some interesting patterns, however.

Figure 13-2. Parameters for Julia Set Pictures

Location	Plate #	Iterations	P	Q	Colors
A	12	512	0.238498	0.512321	17,59,1,0,4,5,20,7, 27,57,58,16,60,61,62,63
B	--	128	0.238498	0.519198	0,16,34,2,18,22,22,8, 58,57,58,59,56,63,63,63
C	13	96	-0.743036	0.113467	0,16,34,2,18,22,23,16' 56,3,35,19,63,2,58,63
D	--	64	-0.192175	0.656734	0,1,17,43,31,63,11,7, 56,57,58,59,60,61,62,63
E	14	32	0.108294	-0.670487	0,46,38,62,54,55,63,54, 46,38,62,54,55,62,62,63
F	--	64	-0.392488	-0.587966	39,1,13,21,47,54,62,7, 56,57,58,59,60,61,62,63
G	--	256	-0.392488	-0.587966	63,1,2,3,4,5,20,8 56,57,58,59,60,61,62,63
H	15	32	0.138341	0.649857	1,32,12,4,36,37,38,32, 12,4,4,38,36,37,39,63
I	16	24	0.278560	-0.003483	25,32,36,54,38,62,63,7 56,57,58,59,60,61,62,63
J	--	24	0.278560	-0.003483	25,32,36,54,38,62,63,7 56,57,58,59,60,61,62,63
K	--	48	-1.258842	0.065330	4,54,51,1,47,44,20,7 56,57,58,59,60,61,62,63
L	--	48	-1.028482	-0.264756	48,51,2,38,4,16,20,7 56,57,58,59,60,61,62,63
M	17	64	0.268545	-0.003483	32,55,36,25,4,5,20,7 56,57,58,59,60,61,62,63
N	--	24	0.268545	-0.003483	32,55,36,25,4,5,20,7 56,57,58,59,60,61,62,63
O	18	256	0.318623	0.044699	8,57,57,59,60,61,62,63, 51,57,58,59,60,61,62,63
P	19	48	0.318623	0.044699	8,57,57,59,60,61,62,63, 51,57,58,59,60,61,62,63

Figure 13-3. Color Selection Data for A

Lower Bound	Upper Bound	Color #1	Color #2
0	6	1	1
6	12	3	4
12	24	5	10
24	50	9	12
50	100	11	14
100	200	12	4
200	512	0	0

Binary Decomposition

We have been drawing Julia sets which show a background color if the iteration of the Julia equation does not blow up to infinity and which cycle through the 16 colors to indicate the iterations required to exceed a threshold when the equation does blow up. Or alternately, we have selected colors for odd and even numbers of iterations and assigned them to a range of iterations.

After we have performed the iterations for each value of the Julia set, there is another piece of information available that we have thus far ignored. This piece of information is the direction of the vector heading for infinity when we are in regions where the function blows up to infinity. You will remember that at each iteration, we obtain a complex number. Furthermore, the test for the function blowing up is that the magnitude of this complex number achieves a value greater than two.

Now, if we consider the real and imaginary parts of the number when that magnitude is achieved, they can be considered as representing a vector in the complex plane which has a direction as well as a magnitude. What we are going to do is determine what this angle is, and color the corresponding point on the Julia picture black if that angle is between 0 and 180 degrees, and white if the angle is between 180 and 360 degrees. (All points that don't blow up have an angle too, namely that of the root which they settle down to, but that information isn't too interesting so we just color it black.) Figure 13-4 is a simplified Julia program to do binary decomposition. You could modify the more detailed Julia program given in Figure 13-1 to do the same thing, or you can use the binary decomposition technique with some of the other curves that will be discussed in future chapters.

Figure 13-5 shows the result of binary decomposition for two Julia sets. The first makes use of a value of 0 for c. If you were to put this value into the Julia set program of Figure 13-1, the result would be rather uninteresting; all that would be produced would be a circle. You can see that much more detail occurs when binary decomposition is performed. The second part of the figure shows binary decomposition of a more typical Julia set. The program of Figure 13-4 includes values of P and Q for both pictures. When the first picture has been completed, hitting the Ent key will blank the screen and start drawing the second picture.

Figure 13-4. Program for Binary Decomposition of Julia Sets

```
program bindecom;

uses CRT,Graph,Fractal;

const
    maxrow: integer = 349;
    maxcol: integer = 639;
    max_iterations: integer = 64;
    max_size: real = 4;

var
    GraphDriver,GraphMode,i,j,row,col,color: integer;
    P,Q,deltaX, deltaY, X, Y, Xsquare,Ysquare,Ytemp,temp1,
    temp2,theta,cos_theta,sin_theta: real;
    ch1: char;

begin

    DirectVideo := false;
    GraphDriver := 4;
    GraphMode := EGAHi;
    InitGraph(graphDriver,GraphMode,'');
    XMax := 2.0;
    XMin := -2.0;
    YMax := 1.5;
    YMin := -1.5;
    P := 0;
```

```
    Q := 0;
    for j:=0 to 1 do
    begin
        deltaX := (XMax - XMin)/(maxcol);
        deltaY := (YMax - YMin)/(maxrow);
        for col:=0 to maxcol do
        begin
            if KeyPressed then
                exit;
            for row:=0 to maxrow do
            begin
                X := XMin + col * deltaX;
                Y := YMax - row * deltaY;
                Xsquare := 0;
                Ysquare := 0;
                i := 0;
                while (i<max_iterations) and ((Xsquare +
                    Ysquare) < max_size) do
                begin
                    Xsquare := X*X;
                    Ysquare := Y*Y;
                    Ytemp := 2*X*Y;
                    X := Xsquare - Ysquare + P;
                    Y := Ytemp + Q;
                    inc(i);
                end;
                if X = 0 then
                    color := 15
                else
                begin
                    cos_theta := abs(X)/(sqrt(X*X + Y*Y));
                    sin_theta := abs(Y)/(sqrt(X*X + Y*Y));
                    theta := arctan (sin_theta/cos_theta);
                    if (X<0) and (Y>=0) then
                        theta := theta + 1.5707963;
                    if (X<0) and (Y<0) then
                        theta := theta + 3.14159625;
                    if (X>0) and (Y<0) then
                        theta := theta + 4.7123889;
                    if (theta>=0) and (theta<=3.14159625)
                        then
                        color := 15
                    else
                        color := 0;
                end;
                PutPixel(col, row, color);
            end;
        end;
        P := 0.318623;
        Q := 0.0429799;
        ch1 := ReadKey;
        ClearDevice;
    end;
end.
```

Figure 13-5. Binary Decomposition of Julia Sets

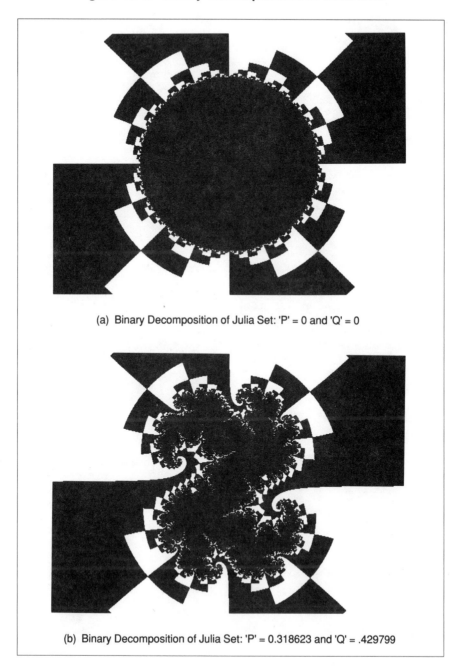

(a) Binary Decomposition of Julia Set: 'P' = 0 and 'Q' = 0

(b) Binary Decomposition of Julia Set: 'P' = 0.318623 and 'Q' = .429799

14

Dragon Curves

The term "dragon curves" covers a lot of territory. Most of these curves were so-named because they looked somewhat like the traditional conception of a mythical dragon. There are several families of curves, however, each with its own distinct mathematical origin. Some of these dragon curves are variations of other curves that we have already dealt with, and should really have been included with them. Nevertheless, we have chosen to put all the dragons together in this chapter.

If you have seen some of the traditional Chinese embroidered dragons, you can note the resemblances as you look at the dragon curves in the color plates and the later pages of this chapter. Some of the prettiest and most dragon-like curves are the "self-squared" dragons first discovered by Mandelbrot. To complicate matters, all of the curves that are produced from variations of the self-squared dragon equation are called dragons, even though some of them do not look anything like dragons, which makes the whole classification very confusing.

Harter-Heightway Dragon

The Harter-Heightway dragon is one of the family of curves created with the initiator/generator technique described in Chapter 6. The first three levels of this curve are shown in Figure 14-1. The first two levels look just like the Polya triangle sweep discussed in Chapter 7. Two additional rules cause this curve to diverge from the Polya triangle so that the dragon curve is generated. They are:

1. The generator alternates between the right and left of the line segment it is replacing.

2. The first position at each level is always to the right.

We use the generic program described in Chapter 6 to generate the dragon curve. However, we must add to the generator function a sign variable which causes the alternation around the line segment being replaced. Figure 14-2 is the program that is used to generate both the Harter-Heightway dragon and the twin dragon. The various levels of detail may be produced by changing the parameter *level*. Figure 14-2 shows the dragon figure produced by setting *level* to 16.

Figure 14-1. First Three Levels of Harter-Heightway Dragon

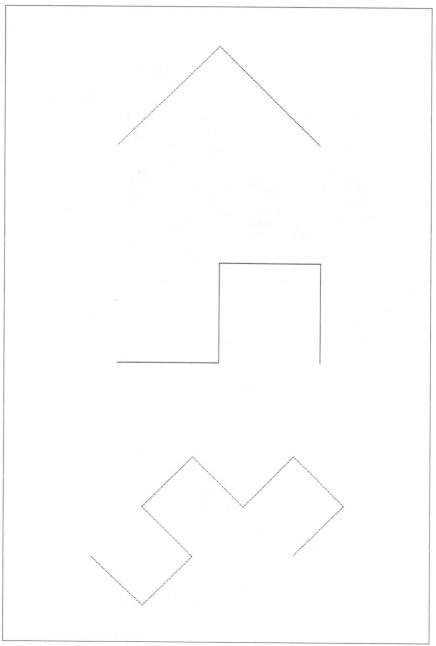

Figure 14-2. Harter-Heightway Dragon

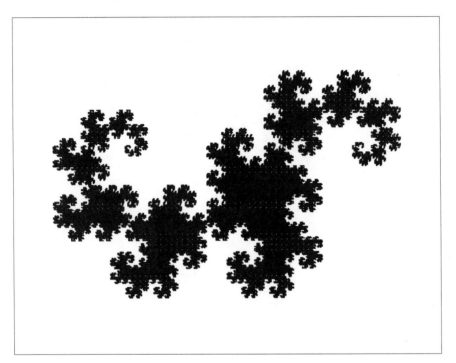

Twin Dragon

It is interesting to see what happens when the same initiator is used, but traversed in the opposite direction. The dragon that is generated fits exactly against the dragon described above. The program of Figure 14-3 is already set up to generate the twin dragon in that it changes color at the end of the *for* loop, and that the proper values for the initiator are given to provide the backward traversing of the original initiator line. For the single dragon, *int_size* needs to be set to 1 instead of 2 so that only one pass is made through the *for* loop and the second set of initiator values is not used. The twin dragon curve is shown in Figure 14-4, using black to depict one dragon and gray to depict the other. It is also shown in color in Plate 20. The exact fit between the two dragons appears in the figure, but to fully appreciate how they mesh together, you need to run the program and watch as the borders of the second dragon are drawn.

Figure 14-3. Program to Generate Harter-Heightway and Twin Dragons

```
program twindrag;

uses CRT,Graph,Fractal;

var
    GraphDriver,GraphMode,i,j,generator_size,level,
        init_size: integer;
    Xpoints, Ypoints: array[0..24] of real;
    initiator_x1,initiator_x2,initiator_y1,initiator_y2:
        array[0..9] of real;
    sign: real;
    ch: char;

procedure generate (X1: real; Y1: real; X2: real; Y2: real;
    level: integer; sign: real);
    var
        j,k: integer;
        sign2: real;
        Xpoints, Ypoints: array[0..24] of real;

    begin
        sign2 := -1;
        turtle_r := (sqrt((X2 - X1)*(X2 - X1) + (Y2 -
            Y1)*(Y2 - Y1)))/1.41421;
        Xpoints[0] := X1;
        Ypoints[0] := Y1;
        Xpoints[2] := X2;
        Ypoints[2] := Y2;
        turtle_theta := point(X1,Y1,X2,Y2);
        turtle_x := X1;
        turtle_y := Y1;
        turn(sign*45);
        step;
        Xpoints[1] := turtle_x;
        Ypoints[1] := turtle_y;
        dec(level);
        if level > 0 then
        begin
            for j:=0 to generator_size - 2 do
            begin
                X1 := Xpoints[j];
                X2 := Xpoints[j+1];
```

```pascal
                Y1 := Ypoints[j];
                Y2 := Ypoints[j+1];
                generate (X1,Y1,X2,Y2,level,sign2);
                sign2 := -sign2;
            end;
        end
        else
        begin
            for k:=0 to generator_size - 2 do
            begin
                    Line(Round(Xpoints[k]+320),Round(175 -
                        Ypoints[k]*0.729)
                        ,Round(Xpoints[k+1]+320),Round(175 -
                        Ypoints[k+1]*0.729));
            end;
        end;
    end;


begin
    GraphDriver := 4;
    GraphMode := EGAHi;
    InitGraph(graphDriver,GraphMode,'');
    SetLineStyle(0,$FFFF,1);
    SetColor(15);
    initiator_x1[0] := -150;
    initiator_x1[1] := 150;
    initiator_x2[0] := 150;
    initiator_x2[1] := -150;
    sign := 1;
    generator_size := 3;
    init_size := 2;
    initiator_y1[0] := -25;
    initiator_y1[1] := -25;
    initiator_y2[0] := -25;
    initiator_y2[1] := -25;
    level := 16;
    for i:=0 to init_size - 1 do
    begin
        generate(initiator_x1[i], initiator_y1[i],
            initiator_x2[i], initiator_y2[i], level,sign);
        SetColor(7);
    end;
    ch := ReadKey;
end.
```

Figure 14-4. Twin Dragon Curves Figure

Julia Dragon

The Julia sets were described in detail in Chapter 13. Depending upon the parameters selected, many different shapes may be created. Some of them, deliberately omitted in Chapter 13, are very dragon-like. You can create a very dragon-like Julia set by running the Julia program using 64 iterations, with palettes set to the colors 42, 32, 36, 54, 38, 62, 63, 32, 36, 54, 38, 62, 63, 32, 36, and 63, and the parameters of $P = 0.383725$ and $Q = 0.147851$.

Self-Squared Dragons

The figures that Mandelbrot calls self-squared dragons result from iterations of the expression:

$$z_{n-1} = cz_n(1-z_n) \qquad \text{(Equation 14-1)}$$

where both z and c are complex numbers, the number c being represented in our program by $c = p + iq$. You will note that if both c and z are real, the equation is the same as that for the population growth curve that we discussed in Chapter 5. As far as I know, no one has looked in detail into the relationship between the bifurcation diagrams and the corresponding dragon curves. This might be a good project for a home computer enthusiast.

Figure 14-5 is the detailed program for generating the self-squared dragon curves. Note that the program is very similar to the program for generating the Julia sets, except for the code that actually performs the iterations of the equation. The program gives you choices for starting a dragon curve from a selected point on the Mandelbrot-like set, for expanding part of an existing dragon display, for completing an unfinished dragon curve drawing, or for quitting. Figure 14-6 lists parameters for various interesting dragon displays. One of these is shown in color in Plate 21 (in the color section).

The data labeled A is the only one that was generated using the color select option. Figure 14-7 lists the color parameters that were selected for that display. As in the case of the Julia sets, increasing the number of iterations increases the detail of the dragon display, but insufficient computer accuracy causes parts of the display to be lost as the iterated values drift away from the attractor. Typically, if we continue to increase the number of iterations, the figure will become fragmented and finally disappear altogether.

Figure 14-5. Program to Generate Self-Squared Dragons

```
program dragon;

uses CRT,Graph,Fractal;

const
    maxcol: integer = 639;
    maxrow: integer = 349;
    max_size: real = 4;
    file_string: array[0..3] of string[7] =
        (('drgset'),('dragon'), ('dragon'),('drgpal'));
    file2: string[5] = '.pcx';
    file3: string[5] = '.pal';
    file_name2: string[20] = 'dragon00.pcx';
    rect: array[1..4] of PointType = ((x: 0; y: 0),
                                      (x: 639; y: 0),
                                      (x: 639; y: 349),
                                      (x: 0; y: 349));

var
    graphDriver,GraphMode,i,j,key,generator_size,level,
        init_size, file_no,color,row,col,error,
        start_col,end_mask, m,color_option,max_iterations:
        integer;
    temp_sq, temp_xy, Ytemp,deltaX,deltaY,X,Y,
        Xsquare,Ysquare,P,Q: real;
    ch1: char;
    colors: array[0..14,0..2] of integer;
    file1: string[3];
    file_name: string[12];
    f: file of byte;
    f_pal: file of integer;

function menu(types: integer): integer;

const
    TextData: array[0..3] of string[45] =
        (('Select Dragon location from "DRGSET" display'),
        ('Expand a Section of an Existing Dragon Curve'),
        ('Complete an Unfinished Dragon Curve'), ('Quit'));
    TextData2: array[0..1] of string[45] =
        (('Select colors and ranges'),
        ('Cycle through colors in order'));

var
    i,k: integer;
    ch1: char;
begin
    k := 0;
```

```
Window(1,1,80,25);
TextBackground(0);
clrscr;
if types = 0 then
begin
    Window(10,6,59,13);
    m := 3;
end
else
begin
    Window(10,6,59,11);
    m := 1;
end;
TextBackground(1);
clrscr;
GotoXY(3,2);
TextColor(10);
write('Make your choice with the up and down arrows:');
repeat
    for i:=0 to m do
    begin
        if i = k then
        begin
            TextColor(1);
            TextBackground(15);
        end
        else
        begin
            TextColor(15);
            TextBackground(1);
        end;
        GotoXY(3,i+4);
        if types = 0 then
            write(TextData[i])
        else
            write(TextData2[i]);
    end;
    ch1:= ReadKey;
    if ch1 = char($00) then
    begin
        ch1 := ReadKey;
        case ch1 of
            'P': begin
                    if k = 3 then
                        k := 0
                    else
                        inc(k);
                end;
            'H': begin
                    if k = 0 then
                        k := 3
```

```
                        else
                            dec(k);
                    end;
            end;
        end;
    until ch1 = char($0D);
    window(1,1,80,25);
    menu := k;
end;

function set_colors: integer;

const
    color_end: integer = 0;
    color_set: array[0..1] of integer = (1,1);
var
    i,k: integer;
    ch1: char;
begin
    TextBackground(4);
    clrscr;
    k := 0;
    Window(10,4,70,22);
    TextBackground(0);
    clrscr;
    GotoXY(3,1);
    TextColor(10);
    writeln('Enter Upper Limit Number and use Arrows to ',
        'Set Colors:');
    writeln;
    TextColor(14);
    write('  Start Iters   End Iters   Color #1  Color #2');
    writeln;
    i := 0;
    colors[0,0] := 0;
    colors[1,0] := 0;
    while colors[i,0] < max_iterations do
    begin
        inc(i);
        TextColor(15);
        GotoXY(6,i+5);
        write(colors[i-1,0]);
        GotoXY(20,i+5);
        Read(colors[i,0]);
        if colors[i,0] <= colors[i-1,0] then
        begin
            colors[i,0] := colors[i-1,0] + 1;
            GotoXY(20,i+5);
            write(colors[i,0]);
        end;
        if (i = 13) or (colors[i,0] > max_iterations) then
```

```
begin
    colors[i,0] := max_iterations;
    GotoXY(20,i+5);
    write(colors[i,0]);
end;
GotoXY(30,i+5);
TextColor(color_set[0]);
write(char($DB),char($DB),char($DB),char($DB),
    char($DB),char($DB));
TextColor(color_set[1]);
GotoXY(40,i+5);
write(char($DB),char($DB),char($DB),char($DB),
    char($DB),char($DB));
for k:= 0 to 1 do
begin
    GotoXY(36+ 10*k,i+5);
    repeat
        ch1:= ReadKey;
        if ch1 = char($00) then
        begin
            ch1 := ReadKey;
            case ch1 of
                'P': begin
                        if color_set[k] = 15 then
                            color_set[k] := 0
                        else
                            inc(color_set[k]);
                    end;
                'H': begin
                        if color_set[k] = 0 then
                            color_set[k] := 15
                        else
                            dec(color_set[k]);
                    end;
            end;
        end;
        GotoXY(30+10*k,i+5);
        TextColor(color_set[k]);
        write(char($DB),char($DB),char($DB),char($DB),
            char($DB),char($DB));
    until ch1 = char($0D);
end;
colors[i,1] := color_set[0];
colors[i,2] := color_set[1];
set_colors := i;
end;
Window(1,1,80,25);
clrscr;
TextColor(15);
end; begin
repeat
```

```
key := menu(0);
if key = 3 then
    exit;
if key < 2 then
begin
    TextBackground(0);
    clrscr;
    Window(15,8,66,11);
    TextBackground(4);
    clrscr;
    GotoXY(3,2);
    TextColor(15);
    write('Enter number of iterations desired',
        '(16-512): ');
    readln(max_iterations);
    if max_iterations < 16 then
        max_iterations := 16;
    if max_iterations > 512 then
        max_iterations := 512;
end;
color_option := menu(1);
if (key < 2) and (color_option = 0) then
    end_mask :=  set_colors;
TextBackground(0);
clrscr;
Window(15,8,66,11);
TextBackground(2);
clrscr;
GotoXY(3,2);
TextColor(15);
if key = 0 then
    write('Enter "DRGSET" file number (00 - 99): ')
else
    write('Enter Dragon Curve number (00 - 99); ');
readln(file_no);
Str(file_no,file1);
Window(1,1,80,25);
if file_no < 10 then
    file_name := Concat
        (file_string[key],'0',file1,file2)
else
    file_name := Concat
        (file_string[key],file1,file2);
    error := restore_screen(file_name);
if error = 0 then
    exit
else
begin
    CURSOR_X := 0;
    CURSOR_Y := 0;
    case key of
```

```
0: begin
     start_col := 0;
     move_cursor(2,15,0,0);
     XMax := 1.4;
     XMin := -0.4;
     YMax := 0.8;
     YMin := -0.8;
     P := Pval;
     Q := Qval;
     SetFillStyle(1,7);
     FillPoly(4,rect);
   end;
2: begin
     if (error = 639) or (error = 0) then
         exit;
     start_col := 8 * (error div 8);
     file_name2 := Copy(file_name,0,
         length(file_name));
     Assign(f,file_name);
     Erase(f);
     P := Pval;
     Q := Qval;
     if color_option = 0 then
     begin
         if file_no < 10 then
             file_name := Concat
                 (file_string[3],'0',file1,file3)
     else
         file_name2 := Concat
             (file_string[3],file1,file3);
         Assign(f_pal,file_name);
         Reset(f_pal);
         for i:= 0 to 14 do
             for j:= 0 to 2 do
                 read(f_pal,colors[i,j]);
         read(f_pal,max_iterations);
         read(f_pal,end_mask);
         Close(f_pal);
     end;
   end;
1: begin
     move_cursor(0,15,0,0);
     move_cursor(1,15,CURSOR_X,CURSOR_Y);
     XMax := TXMax;
     XMin := TXMin;
     YMax := TYMax;
     YMin := TYMin;
     P := Pval;
     Q := Qval;
     start_col := 0;
     SetFillStyle(1,7);
```

```
            FillPoly(4,rect);
        end;
    end;
deltaX := (XMax - XMin)/(maxcol);
deltaY := (YMax - YMin)/(maxrow);
for col:=start_col to maxcol do
begin
    if KeyPressed then
    begin
        Pval := P;
        Qval := Q;
        file1 := save_screen(0,0,col,349,
            file_name2);
        gotoxy(1,24);
        file_name := Concat
            (file_string[2],file1,file2);
        write('File name is: ',file_name);
        if color_option = 0 then
        begin
            file_name := Concat
                (file_string[3],file1,file3);
            Assign(f_pal,file_name);
            Rewrite(f_pal);
            for i:= 0 to 14 do
                for j:= 0 to 2 do
                    write(f_pal,colors[i,j]);
            write(f_pal,max_iterations);
            write(f_pal,end_mask);
            Close(f_pal);
        end;
        exit;
    end;
    for row:=0 to maxrow do
    begin
        X := XMin + col * deltaX;
        Y := YMax - row * deltaY;
        Xsquare := 0;
        Ysquare := 0;
        color := 0;
        while (color<max_iterations) and
            ((Xsquare + Ysquare) < max_size) do
        begin
            Xsquare := X*X;
            Ysquare := Y*Y;
            temp_sq := Ysquare - Xsquare;
            temp_xy := X*Y;
            temp_xy := temp_xy + temp_xy;
            Ytemp := Q*(temp_sq+X)-P*(temp_xy-Y);
             X := P*(temp_sq+X) + Q*(temp_xy-Y);
             Y := Ytemp;
            inc(color);
```

```
                    end;
                    if color_option = 0 then
                    begin
                        i := 0;
                        while (color > colors[i,0]) do
                            inc(i);
                        if color mod 2= 0 then
                            PutPixel(col,row,colors[i][2])
                        else
                            PutPixel(col,row,colors[i][2]);
                    end
                    else
                    begin
                        if color >= max_iterations then
                            color := (Round((Xsquare +
                                Ysquare)*14) mod 14) + 1
                        else
                            color := 0;
                        PutPixel(col,row,color);
                    end;
                end;
            end;
        end;
        Pval := P;
        Qval := Q;
        file1 := save_screen(0,0,639,349,file_name2);
        GotoXY(1,24);
        file_name := ConCat(file_string[2],file1,file2);
        write('File name is: ',file_name);
        ch1 := ReadKey;
        CloseGraph;
    until key = 3;
end.
```

Figure 14-6. Parameters for Dragons

Plate	Iterations	P	Q	Colors
21	256	1.646009	0.967049	0,1,52,62,38,62,52,7, 52,62,38,60,1,63,1,63
A	256	1.335524	0.769341	0,1,2,3,4,5,20,7, 56,57,58,59,60,61,62,63
--	64	1.646009	0.967049	0,1,2,3,4,5,20 56,57,58,59,60,61,62,63
--	64	2.447261	-0.924069	0,62,43,1,9,55,20 56,57,58,59,60,61,62,63
--	64	1.325508	0.786533	8,1,2,3,4,5,20 56,57,58,59,60,61,62,63
--	64	1.415414	0.856803	0,63,62,52,4,5,20 56,57,58,59,60,61,62,63
--	32	1.415414	0.856803	57,24,2,18,10,62,63 56,57,58,59,60,61,62,63
--	128	1.255399	0.691977	0,57,58,59,60,61,63 56,57,58,59,60,61,62,63
--	64	2.797809	-0.657593	8,1,2,3,4,5,20 56,57,58,59,60,61,62,63
--	64	3.018153	-0.098854	5,16,2,38,10,62,36 56,57,58,59,60,61,62,63
22	64	2.998122	0.004298	43,1,2,3,4,5,46,7,57 57,58,59,60,61,62,63 (See text)

(Note: A is not illustrated in color section)

Figure 14-7. Color Selection Data for A

Lower Bound	Upper Bound	Color #1	Color #2
0	5	1	9
5	10	2	10
10	20	3	11
20	30	4	12
30	40	5	13
40	100	6	14
100	200	7	15
200	256	0	0

San Marcos Dragon

Plate 20 shows what Mandelbrot calls the San Marcos Dragon because it reminded him of the San Marcos Square in Venice with its reflection from a wet pavement. To enhance this fantasy, I've used a special color scheme in which everything where y is greater than or equal to zero is colored in brighter shades and everything for y less than zero (the reflection) is colored in darker shades. Figure 14-8 lists the code that must be inserted in the dragon program of Figure 14-5 to perform the proper coloring. This code is inserted in place of the lines that define color for the cycling option just prior to the *PutPixel* statement. Afterwards the *colors* program was used to change the sky and pavement to appropriate shades of blue.

Figure 14-8. Code for Coloring San Marcos Dragon

```
if color >= max_iterations then
    color := (Round((Xsquare + Ysquare)*100) mod 6)+1;
else
    color = 0;
if row < 175 then
    color := color + 8;
```

Dragon Outlines

Before we leave the subject of dragons, it is worth looking at another technique for representing at least the outlines of dragon curves. What is involved, is running the dragon equation backward. Instead of using Equation 14-1, we solve the equation for z_n in terms of z_{n+1}. We then take a representative point and iterate it many times (in this case 12000 times), plotting the location of the point at each iteration. (Each iteration requires taking a square root, which could give either a positive or negative result. To achieve good results from our display, the program is set up to randomly select either the positive or negative square root at each iteration.)

The result of all this is that the point tends to be attracted to the outline of the dragon curve. (The first few points are a little wild before the curve settles down, so we don't plot the first ten points.) This technique doesn't provide the full beauty and detail of the full-fledged dragon curve program, but it is much faster, and is therefore useful in discovering the shape of a dragon before you take the time to make a detailed plot. This also serves as an introduction to techniques that will be gone into in much greater detail in Chapter 20 on iterated function systems. (We have already encountered this method at work in generating the strange attractors in Chapter 4.) Figure 14-9 lists the program to generate dragon outlines. Two typical results of running this program are shown in Figure 14-10.

Figure 14-9. Program to Generate Dragon Outlines

```
program dragout;

uses CRT,Graph,Fractal;

const
    x: real = 0.50001;
    y: real = 0;
    x_center: integer = 320;
    y_center: integer = 175;
var
    col,row,graphDriver,GraphMode,i: integer;
    P,Q,magnitude,scale,temp,temp_x,temp_y: real;
    ch1: char;

begin
    write('Enter P and Q (real and imaginary parameters)',
        separated by a space): ');
    Readln(P,Q);
    magnitude := P*P + Q*Q;
    P := 4*P/magnitude;
    Q := -4*Q/magnitude;
    write('Enter Scale: ');
    Readln(scale);
    scale := x_center*scale;
    GraphDriver := 4;
    GraphMode := EGAHi;
    InitGraph(graphDriver,GraphMode,'');
    for i:=0 to 12000 do
    begin
        temp_x := x*P - y*Q;
```

```
            y := x*Q + y*P;
            temp_y := y;
            x := 1 - temp_x;
            magnitude := sqrt(x*x + y*y);
            y := sqrt((-x + magnitude)/2);
            x := sqrt((x + magnitude) /2);
            if temp_y < 0 then
                x := -x;
            if random < 0.5 then
            begin
                x := -x;
                y := -y;
            end;
            x := (1 - x)/2;
            y := y/2;
            col := Round(scale*(x-0.5) + x_center);
            row := Round(y_center - scale*y);
            if (i > 10) and (col >= 0) and (col < 640) and
                (row >= 0) and (row < 350) then
                PutPixel(col,row,15);
        end;
        ch1 := ReadKey;
end.
```

Figure 14-10. Dragon Outlines

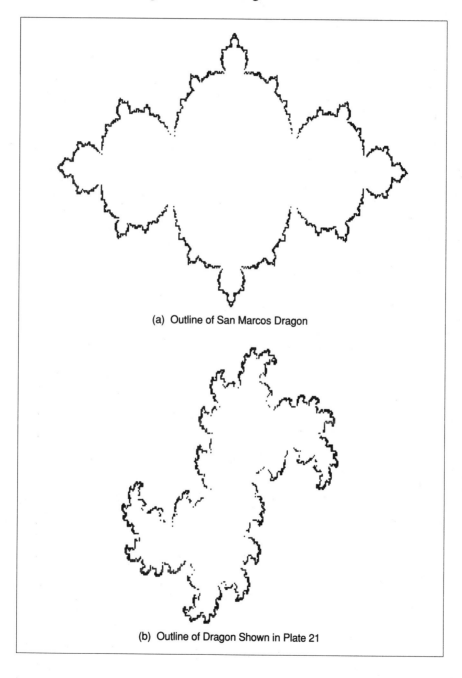

(a) Outline of San Marcos Dragon

(b) Outline of Dragon Shown in Plate 21

Phoenix Curves

The phoenix curve was discovered by Shigehiro Ushiki at Kyoto University. The equations for the phoenix curve are:

$$x_{n+1} = x_n^2 + p + qy_n \qquad \text{(Equation 15-1)}$$

$$y_{n+1} = x_n \qquad \text{(Equation 15-2)}$$

where both x and y are complex. This pair of equations is an easy way to express the relationship, and makes computer handling of it simple, but it does mask somewhat an understanding of what's going on. So let's rewrite the equation in a little more understandable form:

$$x_n = x_{n-1}^2 + p + qx_{n-2} \qquad \text{(Equation 15-3)}$$

Plate 23 (see color section) shows the original phoenix curve. To generate this curve, the values selected for *P* and *Q* are:

$$p = 0.56667 \qquad \text{(Equation 15-4)}$$

$$q = -0.5 \qquad \text{(Equation 15-5)}$$

and 128 iterations are performed. The graph is of *x* in the complex plane, but in order to make the phoenixes stand up correctly, the axes are inverted from normal usage, with the *x* axis representing the imaginary part of *x*, and the *y* axis representing the real part of *x*. Figure 15-1 lists the program to generate phoenix curves. It is quite similar to the programs we used to generate the Julia and dragon curves, but there are some significant differences.

First is the method of coloring the display. This is a good point to re-iterate the methods that we have been using to color our displays. The methods used for the cyclical option for Mandelbrot sets, Julia sets, dragon curves, and the original phoenix curve are summarized in Figure 15-2. The program of Figure 15-1 includes the method which allows you to select colors and ranges, and the method which allows you to cycle through the colors as was done for the Mandelbrot set. The original phoenix curve is generated using the color selecting option. The color values are shown in Figure 15-4.

Figure 15-1. Program to Generate Phoenix Curves

```
program phoenix;

uses CRT,Graph,Fractal;

const
    maxcol: integer = 639;
    maxrow: integer = 349;
    max_size: real = 4;
    file_string: array[0..3] of string[7] =
        (('pheset'),('phenix'),('phenix'),('phepal'));
    file2: string[5] = '.pcx';
    file3: string[5] = '.pal';
    file_name2: string[20] = 'phenix00.pcx';
    rect: array[1..4] of PointType = ((x: 0; y: 0),
                                (x: 639; y: 0),
                                (x: 639; y: 349),
                                (x: 0; y: 349));

    var
        graphDriver,GraphMode,i,j,key,generator_size,level,
            init_size,file_no,color,row,col,error,start_col,
            end_mask,m,color_option,max_iterations: integer;
        deltaX,deltaXi,X,Y,Xi,Yi,Xisquare,Xsquare,Xtemp,
            Xitemp,P,Q: real;
        ch1: char;
        colors: array[0..14,0..2] of integer;
        file1: string[3];      file_name: string[12];
        f: file of byte;
        f_pal: file of integer;
```

```pascal
function menu(types: integer): integer;

const
    TextData: array[0..3] of string[48] =
      (('Select Phoenix Curve location from "PHESET" file'),
       ('Expand a Section of an Existing Phoenix Curve'),
       ('Complete an Unfinished Phoenix Curve'),
       ('Quit'));
    TextData2: array[0..1] of string[45] =
      (('Select colors and ranges'),
       ('Cycle through colors in order'));

var
    i,k: integer;
    ch1: char;
begin
    k := 0;
    Window(1,1,80,25);
    TextBackground(0);
    clrscr;
    if types = 0 then
    begin
        Window(8,6,62,13);
        m := 3;
    end
    else
    begin
        Window(10,6,59,11);
        m := 1;
    end;
    TextBackground(1);
    clrscr;
    GotoXY(3,2);
    TextColor(10);
    write('Make your choice with the up and down arrows:');
    repeat
        for i:=0 to m do
        begin
            if i = k then
            begin
                TextColor(1);
                TextBackground(15);
            end
            else
            begin
                TextColor(15);
                TextBackground(1);
            end;
            GotoXY(3,i+4);
            if types = 0 then
                write(TextData[i])
```

```
            else
                write(TextData2[i]);
        end;
        ch1:= ReadKey;
        if ch1 = char($00) then
        begin
            ch1 := ReadKey;
            case ch1 of
                'P': begin
                        if k = 3 then
                            k := 0
                        else
                            inc(k);
                    end;
                'H': begin
                        if k = 0 then
                            k := 3
                        else
                            dec(k);
                    end;
            end;
        end;
    until ch1 = char($0D);
    window(1,1,80,25);
    menu := k;
end;

function set_colors: integer;

const
    color_set: array[0..1] of integer = (1,1);

var
    i,k: integer;
    ch1: char;
begin
    TextBackground(4);
    clrscr;
        k := 0;
    Window(10,4,70,22);
    TextBackground(0);
    clrscr;
    GotoXY(3,1);
    TextColor(10);
    writeln('Enter Upper Limit Number and use Arrows to ',
            'Set Colors:');
    writeln;
    TextColor(14);
    write('  Start Iters   End Iters   Color #1  Color #2');
    writeln;
    i := 0;
```

```
colors[0,0] := 0;
colors[1,0] := 0;
while colors[i,0] < max_iterations do
begin
    inc(i);
    TextColor(15);
    GotoXY(6,i+5);
    write(colors[i-1,0]);
    GotoXY(20,i+5);
    Read(colors[i,0]);
    if colors[i,0] <= colors[i-1,0] then
    begin
        colors[i,0] := colors[i-1,0] + 1;
        GotoXY(20,i+5);
        write(colors[i,0]);
    end;
    if (i = 13) or (colors[i,0] > max_iterations) then
    begin
        colors[i,0] := max_iterations;
        GotoXY(20,i+5);
        write(colors[i,0]);
    end;
    GotoXY(30,i+5);
    TextColor(color_set[0]);
    write(char($DB),char($DB),char($DB),char($DB),
        char($DB),char($DB));
    TextColor(color_set[1]);
    GotoXY(40,i+5);
    write(char($DB),char($DB),char($DB),char($DB),
        char($DB),char($DB));
    for k:= 0 to 1 do
    begin
        GotoXY(36+ 10*k,i+5);
        repeat
            ch1:= ReadKey;
            if ch1 = char($00) then
            begin
                ch1 := ReadKey;
                case ch1 of
                    'P': begin
                        if color_set[k] = 15 then
                            color_set[k] := 0
                        else
                            inc(color_set[k]);
                        end;
                    'H': begin
                        if color_set[k] = 0 then
                            color_set[k] := 15
                        else
                            dec(color_set[k]);
                        end;
```

```
                        end;
                    end;
                GotoXY(30+10*k,i+5);
                TextColor(color_set[k]);
                write(char($DB),char($DB),char($DB),char($DB),
                    char($DB),char($DB));
            until ch1 = char($0D);
        end;
        colors[i,1] := color_set[0];
        colors[i,2] := color_set[1];
        set_colors := i;
    end;
    Window(1,1,80,25);
    clrscr;
    TextColor(15);
end;

begin
    repeat
        key := menu(0);
        if key = 3 then
            exit;
        if key < 2 then
        begin
            TextBackground(0);
            clrscr;
            Window(15,8,66,11);
            TextBackground(4);
            clrscr;
            GotoXY(3,2);
            TextColor(15);
            write('Enter number of iterations desired,'
            ' (16-512): ');
            readln(max_iterations);
            if max_iterations < 16 then
                max_iterations := 16;
            if max_iterations > 512 then
                max_iterations := 512;
        end;
        color_option := menu(1);
        if (key < 2) and (color_option = 0) then
            end_mask :=  set_colors;
        TextBackground(0);
        clrscr;
        Window(15,8,66,11);
        TextBackground(2);
        clrscr;
        GotoXY(3,2);
        TextColor(15);
        if key = 0 then
            write('Enter "PHESET file number (00 - 99): ')
```

```
else
    write('Enter Phoenix curve number (00 - 99); ');
readln(file_no);
Str(file_no,file1);
Window(1,1,80,25);
if file_no < 10 then
    file_name := Concat
        (file_string[key],'0',file1,file2)
else
    file_name := Concat(file_string[key],
        file1,file2);
error := restore_screen(file_name);
if error = 0 then
    exit
else
begin
    CURSOR_X := 0;
    CURSOR_Y := 0;
    case key of
    0: begin
        start_col := 0;
        move_cursor(2,15,0,0);
        XMax := 1.5;
        XMin := -1.5;
        YMax := 1.2;
        YMin := -1.2;
        P := Pval;
        Q := Qval;
        if color_option = 1 then
        begin
            setEGApalette(0,1);
            setEGApalette(1,57);
            setEGApalette(2,2);
            setEGApalette(3,62);
        end;
        SetFillStyle(1,7);
        FillPoly(4,rect);
    end;
    2: begin
        if (error = 639) or (error = 0) then
            exit;
        start_col := 8 * (error div 8);
        file_name2 := Copy
            (file_name,0,length(file_name));
        Assign(f,file_name);
        Erase(f);
        P := Pval;
        Q := Qval;
        if color_option = 0 then
        begin
            if file_no < 10 then
```

```
                    file_name := Concat
                        (file_string[3],'0',file1,file3)
                else
                    file_name := Concat
                        (file_string[3],file1,file3);
                Assign(f_pal,file_name);
                Reset(f_pal);
                for i:= 0 to 14 do
                    for j:= 0 to 2 do
                        read(f_pal,colors[i,j]);
                read(f_pal,max_iterations);
                read(f_pal,end_mask);
                Close(f_pal);
            end;
        end;
    1: begin
        move_cursor(0,15,0,0);
        move_cursor(1,15,CURSOR_X,CURSOR_Y);
        XMax := TXMax;
        XMin := TXMin;
        YMax := TYMax;
        YMin := TYMin;
        P := Pval;
        Q := Qval;
        start_col := 0;
        SetFillStyle(1,7);
        FillPoly(4,rect);
    end;
    end;
    deltaX := (YMax - YMin)/(maxrow - 1);
    deltaXi := (XMax - XMin)/(maxcol - 1);
    for col:=start_col to maxcol do
    begin
        if KeyPressed then
        begin
            Pval := P;
            Qval := Q;
            file1 :=
                save_screen(0,0,col,349,file_name2);
            gotoxy(1,24);
            file_name := Concat
                (file_string[2],file1,file2);
            write('File name is: ',file_name);
            if color_option = 0 then
            begin
                file_name := Concat
                    (file_string[3],file1,file3);
                Assign(f_pal,file_name);
                Rewrite(f_pal);
                for i:= 0 to 14 do
                    for j:= 0 to 2 do
```

```
                write(f_pal,colors[i,j]);
            write(f_pal,max_iterations);
            write(f_pal,end_mask);
            Close(f_pal);
        end;
        exit;
    end;
    for row:=0 to maxrow do
    begin
        Y  := 0;
        Yi := 0;
        X  := YMax - row * deltaX;
        Xi := XMin + col * deltaXi;
        Xsquare  := 0;
        Xisquare := 0;
        color := 0;
        while (color<max_iterations) and
            ((Xsquare + Xisquare) < max_size) do
        begin
            Xsquare  := X*X;
            Xisquare := Xi*Xi;
            Xtemp  := Xsquare - Xisquare + P + Q*Y;
            Xitemp := 2*X*Xi + Q*Yi;
            Y  := X;
            Yi := Xi;
            X  := Xtemp;
            Xi := Xitemp;
            inc(color);
        end;
        if color_option = 0 then
        begin
            i:=0;
            while (color > colors[i,0]) do
                inc(i);
            if color mod 2= 0 then
                PutPixel(col,row,colors[i][2])
            else
                PutPixel(col,row,colors[i][1]);
        end
        else
        begin
            if color >= max_iterations then
                color := 0
            else
                color := color mod 14 + 1;
            PutPixel(col,row,color);
        end;
    end;
end;
Pval := P;
Qval := Q;
```

```
            file1 := save_screen(0,0,639,349,file_name2);
            gotoxy(1,24);
            file_name := ConCat(file_string[2],file1,file2);
            write('File name is: ',file_name);
            ch1 := ReadKey;
            CloseGraph;
        end;
    until key = 3;
end.
```

Figure 15-2. Coloring Techniques

Type of Figure	Background Color	Other Colors
Mandelbrot Set	All points that do not blow up during all iterations.	Cycle through 15 other colors for number of iterations required for blow-up.
Julia Set	All points that blow up to infinity during iterations.	Cycle through 6 other colors for ranges of values of magnitude of result after iterations are complete.
Original Phoenix Curve	All points that do not blow up during all iterations.	Three colors: one for blow-up in 1 to 32 iterations; second for blow-up in 33 to 64 iterations; third for blow-up in more than 64 iterations.

Figure 15-3. Parameters of Phoenix Curves

Plate #	P	Q
23	0.56667	-0.50000
--	0.288732	0.510029
--	0.484194	-1.044413
24	0.356338	-1.209169
--	-0.550704	-1.255014

Figure 15-4. Color Selection Data for Original Phoenix Curve

Lower Bound	Upper Bound	Color #1	Color #2
0	32	1	1
32	64	9	9
64	126	2	2
126	128	14	14

Maps of the Phoenix Curves

In Chapter 12 we provided a program for generating a set which was the equivalent of the Mandelbrot set for phoenix curves. In previous chapters we have used the Mandelbrot set as a map to permit us to select likely locations for interesting Julia sets, and we have used the equivalent of the Mandelbrot set for dragon curves to perform a similar function in selecting likely dragon curves. The use of the set that acts as a map of the phoenix curves works just fine if we start with initial values for X, Xi, Y, and Yi which cause the derivative of the iterated expression to be zero.

However, we must also consider that we have used the parameters $XMin$, $YMin$, $XMax$, and $YMax$ to save the minimum and maximum values corresponding to the start of the x and y axes and the end of the x and y axes respectively, and that we have saved these values in the screen files. For the Mandelbrot and Julia sets, and for their equivalents for the dragon curves, the $XMin$ and $XMax$ values corresponded either to the real part of a parameter of the equation or to the real part of the initial value of the function being iterated. Similarly, the values of $YMin$ and $YMax$ represented the values of the imaginary part. For the phoenix equivalent of the Mandelbrot set, there are two parameters and we always take them as being real numbers. (There is no reason why this has to be true; one could make both parameters complex numbers, resulting in a function that was a lot more complicated, but might possibly result in some astounding new and previously undiscovered curves.)

At any rate, the set is plotted with P on the x axis and Q on the y axis, but there is no correspondence between the real and imaginary parts of any parameter and the x and y axes for this curve. When we come to the program which generates the phoenix curves themselves, the plotting is such that the real part of the variable x is plotted on the y axis, and the imaginary part of x is plotted on the x axis. This requires that the program do a little juggling when it retrieves the $XMin$, etc., values and the P and Q parameters for a screen file, in order to put them into the proper parameters for the program to properly generate the phoenix curve.

With the Julia and dragon curves, choosing values of P and Q that corresponded to distinct cusps on the associated sets resulted in interesting and beautiful curves. The cusps are less prominent on the map of phoenix curves; in fact the original phoenix curve corresponds to a fairly smooth portion of the set boundary. The values are also quite critical for the most interesting displays, although you can produce a curve quite like the original phoenix curve by coming as close to the values give in Equations 15-4 and 15-5 as you can get with the cursor, using the program.

Plate 23 shows the original phoenix curve and Plate 24 shows a phoenix curve from a different location (see color section). This second curve is generated using the color cycling option. Figure 15-3 shows the parameters for these curves and for some other phoenix curves that you might want to try as a beginning.

16

Three-Dimensional Dragons

Some of the most interesting fractal curves have been generated by Alan Norton at IBM's Thomas J. Watson Research Center. Norton specializes in expressing the dragon equation in multi-dimensional form, using quaternions. His three-dimensional dragon pictures are not only beautiful, but provide an eerie sense of familiarity, no doubt because fractals are the language of natural things. Norton uses IBM's computer resources to plot a million or more points in three dimensions to generate a three-dimensional dragon curve. He then runs a ray tracing program which determines the illumination of every point and the positioning of it on a two-dimensional display. Needless to say, this is beyond the capability of our personal computers.

Nevertheless, we don't want to give up on three-dimensional dragons. The program given below will make use of the dragon outline program given in Chapter 14 to draw cross-sections of the three-dimensional dragon outline. Using P as the third dimension, it will calculate repeated dragon outlines and project them from three-dimensional to two-dimensional space. The results are less complete than Norton's, but give the appearance of three-dimensional displays.

Method of Projection

Suppose we have a program that generates all of the coordinate information needed to create an object in three-dimensional space. Our problem is that we somehow want to display this information on the two-dimensional display screen. The three-dimensional space is defined by the coordinate system (x,y,z) centered at $(0,0,0)$. The display screen is defined by the coordinate system (v_x,v_y) with its origin at $(0,0)$.

First, we place the three-dimensional space so that its origin coincides with the origin of the display screen. Now, each of the axes of the three-dimensional coordinate system makes an angle with the display plane. We shall call the angle between the x axis and the display plane Γ, the angle between the y axis and the display plane α, and the angle between the z axis and the display plane β. Projection of the three-dimensional space onto the display plane now becomes very simple. There is only one line that can be drawn perpendicular to the display plane which intersects any particular point (x,y,z) in the three-dimensional space. The point where this line intersects the display plane is the projection of the three-dimensional point onto the display plane. The equations for making this projection are:

$$v_x = x \cos \Gamma + y \cos \alpha + z \cos \beta \qquad \text{(Equation 16-1)}$$

$$v_y = x \sin \Gamma + y \sin \alpha + z \sin \beta \qquad \text{(Equation 16-2)}$$

While you can use any angles that you desire for Γ, α, and β, some orientations are not very interesting. You have to be careful not to orient the three-dimensional figure so that the two-dimensional projection of it is just a flat on view of one side, with all information about the third dimension lost. The most common angles that are used are those for isometric drawing, where Γ and α are thirty degrees and β is ninety degrees; and the orientation used in many geometry books, where Γ and α are zero and β is 135 degrees.

Programming the Three-Dimensional Dragon

Figure 16-1 is a listing of a program to generate three-dimensional dragons. You will recognize the program from Chapter 14, which draws the outlines of dragons as the heart of this program. We use the same two dimensions as in Chapter 14 for two of the three dimensions of our dragon, and the parameter P (the real part of the multiplier) as the third dimension. The program allows you to enter the projection angles, the scale, and the parameter Q (the imaginary part of the multiplier), and an x and y offset.

We reiterate the outline drawing program over a range of values of P, but for each point that is calculated we use the projection equations given in the section above to determine where to plot the point on the display screen, instead of plotting it directly. The result is a three-dimensional dragon image. The weakness of this technique for drawing three-dimensional figures is that it does not provide cues for separating out hidden surfaces and eliminating them, and it has no way to determine the illumination that should be associated with each point.

The first is not a major objection. We see the dragon as a sort of semi-transparent entity, in which we see the hidden parts of the figure through the parts that are in front. This gives us a better understanding of the geometry involved, and is not a great hardship since semi-transparent dragons are certainly within the realm of possibility. As for the lighting problem, the program makes a crude attempt to alleviate this by using dark colors to paint points of rearward surfaces and light colors to paint the forward ones.

The program provides for two computation options. If you enter a value of zero for Q, the program computes values over a range for the P value of -1 to + 1. As the program steps through various values of P, it holds the magnitude of $P + iQ$ constant at one, so that values are taken around a unit circle. Plate 25 is an example of this mode of the program in action. The angles that were used are Γ = 30 degrees, α = -

30 degrees, and β = 90 degrees. The scale factor was 0.20, the x displacement was zero, and the y displacement was +10. Needless to say, the value entered for Q was 0.

The other mode of operation occurs when a value of Q other than zero is entered. The program then holds the value of Q constant and steps through a range of values of P from -3.0 to +3.0. Plate 26 is an example of this mode of operation. For this picture, the angles were Γ = 15 degrees, α = -15 degrees, and β = 90 degrees. The scale factor was 0.3 and the value entered for Q was 0.967049. No offsets were used. These examples give you some good starting points for your own investigations. Be warned, however, that it easy to select unfortunate combinations of values which result in nothing being displayed at all because all points are beyond the display limits. The best thing to do is use a very small scale factor (0.1 or less) for your first attempt. This will show you what the display looks like and give you some idea of the scale factor and offsets required for a well centered final display.

Figure 16-1. Program to Draw Three-Dimensional Dragons

```pascal
program gen3ddrg;

uses CRT,Graph,Fractal;

const
    x_center: integer = 320;
    y_center: integer = 175;
    x: real = 0.50001;
    y: real = 0;
    step_size: real = 0.4;
    upper_limit: real = 3.0;
    lower_limit: real = -3.0;
    file_name: string[14] ='3ddrag00.pcx';

var

    GraphDriver,GraphMode,i,j,row,col,color1,color: integer;
    z,P,Q,k,sx,cx,sy,cy,sz,cz,magnitude,scale,temp,
        temp_x,temp_y,ymax,ymin,alpha,beta,gamma,QVal,
        x_offset, y_offset: real;
    file_no: string[3];
    ch1: char;
```

```
procedure projection(x3: real; y3: real; z3: real);
    var
        temp_x, temp_y: real;
    begin
        temp_x := x3*cx + y3*cy + z3*cz;
        temp_y := x3*sx + y3*sy + z3*sz;
        col := Round(scale * (temp_x-0.5) + x_center +
            x_offset);
        row := Round(y_center - scale*temp_y + y_offset);
        color := Round(abs(y3*7)) mod 7 + 1;
        if y3>0 then
            color := color + 8;
        if (col>=0) and (col<640) and (row>=0) and (row<350)
            then
            PutPixel(col,row,color);
    end;  function degrees_to_radians(degrees: real): real;
    const
        rad_per_degree: real = 0.0174533;

    var
        angle: real;

    begin
        while degrees >= 360 do
            degrees := degrees - 360;
        while degrees < 0 do
            degrees := degrees + 360;
        angle := rad_per_degree*degrees;
        degrees_to_radians := angle;
    end;

begin
    DirectVideo := false;
    ClrScr;
    write('Enter alpha: ');
    Readln(alpha);
    write('Enter beta: ');
    Readln(beta);
    write('Enter gamma: ');
    Readln(gamma);
    write('Enter scale: ');
    Readln(scale);
    scale := x_center * scale;
    write('Enter X offset: ');
    Readln(x_offset);
    write('Enter Y offset: ');
    Readln(y_offset);
    write('Enter Q parameter: ');
    Readln(QVal);
    if QVal = 0 then
    begin
```

```
        step_size := 0.1;
        upper_limit := 1.0;
        lower_limit := -1.0;
    end;
    GraphDriver := 4;
    GraphMode := EGAHi;
    InitGraph(graphDriver,GraphMode,'');
    alpha := degrees_to_radians(alpha);
    sx := sin(alpha);
    cx := cos(alpha);
    beta := degrees_to_radians(beta);
    sy := sin(beta);
    cy := cos(beta);
    gamma := degrees_to_radians(gamma);
    sz := sin(gamma);
    cz := cos(gamma);
    color := 1;
    k:= upper_limit;
    while k >= lower_limit do
    begin
        if (k<1.0) and (k>-1.0) then
            step_size := 0.1;
        x := 0.50001;
        y := 0;
        if QVal = 0 then
        begin
            magnitude := 1;
            Q := 4*sqrt(1-k*k);
        end
        else
        begin
            magnitude := k*k + QVal*QVal;
            Q := -4*QVal/magnitude;
        end;
        P := 4*k/magnitude;
        for i:=0 to 12000 do
        begin
            temp_x := x*P - y*Q;
            y := x*Q + y*P;
            temp_y := y;
            x := 1 - temp_x;
            magnitude := sqrt(x*x + y*y);
            y := sqrt((-x + magnitude)/2);
            x := sqrt((x + magnitude) /2);
            if temp_y < 0 then
                x := -x;
            if (random < 0.5) then
            begin
                x := -x;
                y := -y;
            end;
```

```
            x := (1 - x)/2;
            y := y/2;
            z :=  P/2;
            if i>10 then
                projection (x, y, z);
        end;
        k := k - step_size;
    end;
    file_no := save_screen(0,0,639,349, file_name);
    ch1 := ReadKey;
end.
```

17

Newton's Method

Newton's method is an iterated numerical approximation technique developed by Sir Isaac Newton to obtain the solutions of equations that do not have a closed form solution. The method works as follows:

1. Suppose we have the generalized equation:

    $$f(z) = 0 \qquad \text{(Equation 17-1)}$$

    where z may be a complex number.

2. Make a guess as to a root of this equation. Call the guess z_0.

3. Compute the expression:

    $$z_{n+1} = z_n - (f(z_n))/(f'(z_n)) \qquad \text{(Equation 17-2)}$$

    where f' is the derivative of f, and we start with z_0 on the right side of the equation to obtain z_1 on the left side.

4. Now repeat this process as many times as you desire. Each new value for z will be a closer approximation to the root of the equation.

Observe that this iteration process is very similar to that which we have used in previous chapters to obtain the Julia, dragon, and phoenix curves. Thus, we have opened the door for plotting curves for a very wide family of equations, and for obtaining graphical results which have a mathematical meaning.

Programs for Plotting Newton's Method Curves

We shall provide programs for plotting the curves of two Newton's method equation solutions. From there on, you're on your own as to setting up similar programs for any equations you can dream up. Figure 17-1 is the listing of a program to obtain the Newton's method solution of the equation:

$$z^3 - 1 = 0 \qquad\qquad \text{(Equation 17-3)}$$

Actually, one thing in the listing of Figure 17-1 is not very good programming practice. This is the use of an if statement which compares pairs of non-integers (the old and new values of x and y) to determine whether to exit the while loop before the maximum number of iterations takes place. Two of the roots of the equation are transcendental numbers; if the precision of the computer were infinite, each iteration would increase the accuracy of these roots, but the old and new values would never match.

Thus, we would perform the maximum number of iterations each time, resulting in those points that zeroed in on these two roots being displayed in only two colors, for a very uninteresting display. Actually we get a good display because the precision of the computer is limited so that after a certain number of iterations the amount to be added to the previous root computation is zero, and the comparison is successful. Unfortunately, this is more good luck than good sense.

We next attempt the same procedure for:

$$z^3 - 2z - 5 = 0 \qquad\qquad \text{(Equation 17-4)}$$

Now, our luck runs out, since at the limit of the computer precision, the previous root computation has its least significant decimal place, alternately incremented and decremented for as long as computations take place. The test is never successful, so we iterate the maximum number of times and produce the same uninteresting display that would occur for infinite precision. To get around this, we set up *xnew* and *ynew* which are values of the current *x* and *y* and *xold* and *yold* which are the previous values of *x* and *y*. These parameters are real numbers, whereas our computations are double precision numbers. We make the comparison with the real numbers; any vagaries of the least significant digits of the original computations are truncated and the test is therefore successful.

Don't try to port this program to a different type of computer or a different version of Pascal unless you are sure how real and double precision numbers are defined in the new environment. The resulting program is listed in Figure 17-2.

Figure 17-1. Program to Solve $z^3 - 1 = 0$

```
program newton3;

uses CRT,Graph,Fractal;

const
    maxcol: integer = 639;
    maxrow: integer = 349;
    max_colors: integer = 16;
    max_iterations: integer = 64;
    max_size: integer = 4;
    color_set: array[0..15] of integer = (1,9,25,15,27,48,2,
        18,62,63,4,37,60,46,45,47);
    file_name: string[15] = 'newton00.pcx';
var
    flag,GraphDriver,GraphMode,color,col,row,i: integer;
    deltaX, deltaY, X, Y, Xsquare,Xold,Yold,
        Ysquare,Ytemp,temp1,temp2,denom,theta: double;
    ch1: char;
    file_no: string[3];

begin
    DirectVideo := false;
    GraphDriver := 4;
```

```
GraphMode := EGAHi;
InitGraph(graphDriver,GraphMode,'');
for i:= 0 to 15 do
    setEGApalette(i,color_set[i]);
XMax := 3.5;
XMin := -3.5;
YMax := 2.50;
YMin := -2.50;
deltaX := (XMax - XMin)/(maxcol);
deltaY := (YMax - YMin)/(maxrow);
for col:=0 to maxcol do
begin
    if KeyPressed then
        exit;
    for row:=0 to maxrow do
    begin                   X := XMin + col * deltaX;
        Y := YMax - row * deltaY;
        Xsquare := 0;
        Ysquare := 0;
        Xold := 42;
        Yold := 42;
        i := 0;
        flag := 0;
        while (i <= max_iterations) and (flag = 0) do
        begin
            Xsquare := X*X;
            Ysquare := Y*Y;
            denom := 3*((Xsquare - Ysquare)*(Xsquare
                - Ysquare) + 4*Xsquare*Ysquare);
            if denom = 0 then
                denom := 0.00000001;
            X := 0.6666667*X + (Xsquare -
                Ysquare)/denom;
             Y := 0.6666667*Y - 2*X*Y/denom;
            if (Xold = X) and (Yold = Y) then
                flag := 1;
            Xold := X;
            Yold := Y;
            inc(i);
        end;
        if X>0 then
            color := i mod 5
        else
        begin
            if (X<-0.3) and (Y>0) then
                color := (i mod 5) + 5
            else
                color := (i mod 6) + 10;
        end;
        PutPixel(col, row, color);
    end;
```

```
    end;
    file_no := save_screen(0,0,639,349, file_name);
    ch1 := ReadKey;
end.
```

Figure 17-2 Program to Solve $z^3 - 2z - 5 = 0$

```
program newton3i;

uses CRT,Graph,Fractal;

const
    maxcol: integer = 639;
    maxrow: integer = 349;
    max_colors: integer = 16;
    max_iterations: integer = 64;
    max_size: integer = 4;
    color_set: array[0..15] of integer = 1,8,25,27,11,4,36,
        52,61,45,2,16,23,35,62,63);
    file_name: string[15] = 'newton00.pcx';
var
    flag,GraphDriver,GraphMode,color,col,row,i: integer;
        deltaX, deltaY,X,Y,Xsquare,Ysquare,Ytemp,
        temp1,temp2, temp3,denom,theta: double;
    xold,yold,xnew,ynew: real;
    ch1: char;
    file_no: string[3];

begin
    DirectVideo := false;
    GraphDriver := 4;
    GraphMode := EGAHi;
    InitGraph(graphDriver,GraphMode,'');
    for i:= 0 to 15 do
        setEGApalette(i,color_set[i]);
    XMax := 3.5;
    XMin := -3.5;
    YMax := 2.50;
    YMin := -2.50;
    deltaX := (XMax - XMin)/(maxcol);
    deltaY := (YMax - YMin)/(maxrow);
    for col:=0 to maxcol do
    begin
        if KeyPressed then
            exit;
        for row:=0 to maxrow do
        begin
            X := XMin + col * deltaX;
```

```
            Y := YMax - row * deltaY;
            Xsquare := 0;
            Ysquare := 0;
            Xold := 42;
            Yold := 42;
            i := 0;
            flag := 0;
            while (i <= max_iterations) and (flag = 0) do
            begin
                Xsquare := X*X;
                Ysquare := Y*Y;
                denom := (3*Xsquare - 3*Ysquare - 2);
                denom := denom*denom + 36*Xsquare*Ysquare;
                if denom = 0 then
                    denom := 0.00000001;
                temp1 := X*Xsquare - 3*X*Ysquare - 2*X -5;
                temp2 := 3*Xsquare - 3*Ysquare - 2;
                temp3 := 3*Xsquare*Y - Ysquare*Y - 2*Y;
                X := X - (temp1 * temp2 -
                    6*X*Y*temp3)/denom;
                Y := Y - (temp1 * (-6*X*Y) + temp3 *
                    temp2)/denom;
                Xnew := X;
                Ynew := Y;
                if (Xold = Xnew) and (Yold = Ynew) then
                    flag := 1;
                Xold := X;
                Yold := Y;
                inc(i);
            end;
            if X>0 then
                color := i mod 5
            else
            begin
                if (X<-0.3) and (Y>0) then
                    color := (i mod 5) + 5
                else
                    color := (i mod 6) + 10;
            end;
            PutPixel(col, row, color);
        end;
    end;
    file_no := save_screen(0,0,639,349, file_name);
    ch1 := ReadKey;
end.
```

Mathematical Meaning of the Curves

Plate 27 (see color section) shows the resulting curve from running the program of Figure 17-1. If you will look at the program listing, you will see that the colors are grouped in sets of 5, 5, and 6, respectively. Since Newton's method always settles on one of the roots of the equation, we don't need to worry about the expression blowing up to infinity.

What we have done, is provide three groups of colors, one for each of the three roots of the equation to which the solution may be attracted. You need some prior knowledge of where the roots are located in order to do this. One way to get this information is to run the program for a little while and print out the resulting root. This is pretty simple for the equation $z^3 - 1 = 0$, since one root is obviously 1. Dividing the equation by $z - 1$ leaves a quadratic equation which can be solved by the quadratic formula to obtain roots of $-.5 \pm .8666i$. Knowing that the ultimate value of z will always be one of these three values, we can devise some if tests to determine which color group to use.

We originally used *colors* to change the 16 default colors so that the group that is attracted toward the root 1 is shades of blue, the group for $-.5 + .866i$ is shades of green, and the group for $-.5 - .8666i$ is shades of red. The resulting colors were written into the program so that you won't need to do any color shifting. Plate 28 (see color section) shows the resulting curve from running the program of Figure 17-2. The roots of this equation are different from the roots given above, but once they were determined, it was observed that the same tests used for the first program could differentiate between the three sets of roots.

Note that for either program, the different shades of a basic color represent the number of iterations that were required for the program to converge to a root. The test for the first program is simply that the same x and y values occurred on the current iteration as occurred for the previous one. For the second program, we need to use double precision numbers for computation and single precision (real) numbers for comparisons, as explained above. Now it is time to consider what these curves mean.

Any point on the figure represents a complex number which can be used as a starting point for the Newton's method process. It was once assumed that you started by making a good guess as to the value of a root, and that Newton's method would then converge to the nearest root. The two figures show very plainly that this is not true. There are lots of color mixtures and isolated islands of one color within another, which demonstrate that it is quite possible to pick an initial value that converges to a root quite far away from the initial value.

Note particularly in Plate 28 the mixed color areas on the left side of the figure. In these areas, a very tiny change in the initial value selected will result in convergence to a totally different root. Another thing that is of interest is the path which the variable takes on its way to convergence. We have not attempted to plot this, but it appears to be very interesting. Note that if we take one of the initial tiny points of green, for example, on the left side of Plate 28, it must land on a green point at every step in the convergence process. If it landed on another color, the succeeding iterations would be just the same as if we had begun at that point, so convergence would be to a different root, which is a contradiction of terms.

18

Brownian Motion

Up until now we have been looking at fractals that are deterministic-ally defined by relatively simple iterated equations. It is tempting to think that by using these techniques, we can mathematically describe (and consequently, realistically represent) any natural phenomena. But no matter how much rich and apparently irregular detail appears in fractal curves, they are never exactly like natural phenomena, which is filled with random irregularities. Your body may have been generated by the regular fractal patterns defined by your genetic structure, but the scar on your arm that you got when you cut yourself when you were twelve years old is a random departure from the fractal pattern that cannot be covered by the mathematical expressions. Nature is full of these random departures from regularity, and to properly represent them, we need to look at introducing some randomness into our fractal techniques.

The first detailed investigation of such randomness in nature was conducted by Robert Brown in 1828. Brown was looking at the movement of pollen particles. This movement had irregularities for which there was no mathematical explanation. Similar movements have since been observed in the dispersion of minute particles through a fluid and in the mixture of different colored gases. Investigations of this random movement, called Brownian motion after its discoverer, have suggested that it occurs in particles so small that collisions on a molecular level are not regular from all directions, so that the unsymmetric application of kinetic molecular forces causes the particles to be propelled in random direction and with random velocities over time.

This chapter will introduce techniques for generating representations of Brownian motion. The resulting curves are not too interesting in themselves, but the tools used here are powerful in generating natural scenes through fractal means. Unfortunately, in the next chapter, where we provide some scene-drawing techniques, we have not been able to make too much use of the randomized methods. They do appear in one program, to draw a scene of earth viewed from the moon, but at the level of computation required to be compatible with the limitations of personal computers, they do not make a significant difference. They do play a significant part in creating some of the breathtaking scenes that have been produced on mainframe computers; consequently, this introduction to the subject, is included so that you can achieve some familiarity with the subject and so that you will have the tools for your own experimentation.

One-Dimensional Brownian Motion

The simplest form of Brownian motion occurs when time is divided into units and plotted along the x axis, and the length traveled by a particle in that time interval is plotted on the y axis. For this case, the y values have been shown to have a Gaussian distribution. Figure 18-1 lists a program to plot the simple Brownian motion. The program begins with determining a random value of the Brownian function at the beginning and end of the plot. This determines the y value of the function at the endpoints.

If you want to take some sort of random curve, defined by line segments, and replace the line segments by Brownian curves, you need to set the beginning and end points for each line segment to match the beginning and end coordinates of the line to assure continuity of the function from one line segment to the next. Alternately, you can use random settings for the beginning and end points of the first line segment, then use the value that you obtained for the end point of that segment as the beginning point for the next line segment and find a random end point for that line segment, and so forth.

Before proceeding further, let's look at the function *gauss*, which returns a random number that has a gaussian distribution. I am indebted to R.A.J. Taylor of Ohio State University for calling to my attention the summation method used here to obtain a Gaussian Distribution. This function begins by looking at a "seed" parameter. If this parameter is some number other than zero, the function reseeds the random number generator; if the seed is zero, the random number generator proceeds to generate random numbers without being restarted. The random number generator returns real numbers between 0 and 1.0. First, twelve of these are summed. Six is then subtracted from the sum to give a Gaussian random number between -6 and +6. This number is multiplied by the standard deviation and the mean is added to it to obtain the final result.

The heart of this program is a procedure called *subdivide*. On the first pass, what this function does is divide the line connecting the beginning and end points at the center, and then use *gauss* to obtain a random value of the gaussian distribution, which is multiplied by a scale factor and stored in an array as the *y* coordinate of the brownian function at that point. It then calls itself recursively twice to perform the same procedure with each half of the line, reducing the scale factor for these new iterations by a specified ratio. This recursive process continues until the computed midpoint is the same as one of the ends of the line. Since the line coordinates are integer values, this means that the array of points has been filled so the function terminates.

The main program, after determining the beginning and end points, establishes the scale factor and reduction ratio that will be used, then calls *subdivide* to fill the array of points, and then utilizes a *for* loop to draw a line between each pair of adjacent points. Before each iteration of this process you are given the opportunity to enter a seed

for the random number generator. Each different seed results in a
unique Brownian curve. If you enter the seed 0, the curve for a zero
seed will be plotted, but upon the next entry of a character on the
keyboard, the program will terminate. Figure 18-2 shows several
representations of Brownian motion obtained from this program.

Figure 18-1. Program to Generate Brownian Fractals

```
program brownian;

uses CRT,Graph,Fractal;

const
    scale: real = 80;
    h: real = 0.87;
    seed: longint = 3456;
    mu: real = 0.0;
    sigma: real = 1.0;
var
    Fh: array[0..256] of real;
    ratio,std: real;
    GraphDriver,GraphMode,i: integer;
    ch1: char;

function gauss(seed: longint): real;

var
    value,exponent,gaussian: real;

begin
    if seed <> 0 then
        RandSeed := seed;
    x: = 0;
    for i: = 0 to 11 do
        x: = x + Random;
    x: = x - 6.0;
    gauss : = mu + sigma*x;
end;

procedure subdivide (f1: integer; f2: integer; std: real);

    var
        fmid: integer;
        stdmid: real;

    begin
        fmid := (f1 + f2) div 2;
        if (fmid <> f1) and (fmid <> f2) then
```

```
            begin
                Fh[fmid] := (Fh[f1] + Fh[f2])/2.0 + gauss(0)
                    * std;
                stdmid := std*ratio;
                subdivide(f1,fmid,stdmid);
                subdivide(fmid,f2,stdmid);
            end;
        end;

begin
    while seed <> 0 do
    begin
        CloseGraph;
        write('Enter seed (0 to quit): ');
        Readln(seed);
        GraphDriver := 4;
        GraphMode := EGAHi;
        InitGraph(graphDriver,GraphMode,'');
        SetLineStyle(0,$FFFF,1);
        SetColor(15);
        Fh[0] := gauss(seed) * scale;
        Fh[256] := gauss(0) * scale;
        ratio := Exp(-0.693147*h);
        std := scale*ratio;
        subdivide(0,256,std);
        for i:=0 to 255 do
            Line(2*i+80,175-Round(Fh[i]),2*(i+1)+80,
                175-Round(Fh[i+1]));
        SetLineStyle(3,$0F0F,1);
            Line(80,175,590,175);
        SetLineStyle(0,$FFFF,1);
        ch1 := ReadKey;
    end;
end.
```

Figure 18-2. Typical One-Dimensional Brownian Motion

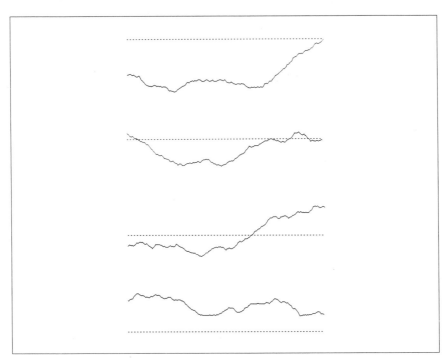

Two-Dimensional Brownian Motion

The above program was confined to describing particle motion in a single dimension. If we now expand to consider the motion of a Brownian particle on a plane, we have two methods of graphing our result. The first technique makes use of a three-dimensional graph and shows particle position in two dimensions plotted against time. The second method uses a two-dimensional graph and simply plots the path of the particle without reference to the time involved. We shall use this second technique. Figure 18-3 lists a program to produce a graph of Brownian motion using this technique. It is essentially the same as the program listed in Figure 18-1, except that we have increased the number of points to be computed, changed the scale, and provided for calculating two dimensions for each point and saving these values in two arrays. The program then draws a line between each two adjacent sets of *x* and *y* coordinates. Figure 18-4 shows several resulting Brownian motion curves obtained using this program.

Figure 18-3. Program to Generate Two Dimensional Brownian Fractals

```
program brown2d;

uses CRT,Graph,Fractal;

const
    scale: real = 100;
    h: real = 0.87;
    seed: longint = 3045;

var
    Fh: array[0..2048] of real;
    Fw: array[0..2048] of real;
    ratio,std: real;
    GraphDriver,GraphMode,i: integer;
    ch1: char;

function gauss(seed: longint): real;

var
    value,exponent,gaussian: real;

begin
    if seed <> 0 then
        RandSeed := seed;
    x: = 0;
    for i: = 0 to 11 do
        x: = x + Random;
    x: = x - 6.0;
    gauss : = mu + sigma*x;
end;

procedure subdivide (f1: integer; f2: integer; std: real);
    var
        fmid: integer;
        stdmid: real;

    begin
        fmid := (f1 + f2) div 2;
        if (fmid <> f1) and (fmid <> f2) then
        begin
            Fh[fmid] := (Fh[f1] + Fh[f2])/2.0 + gauss(0)
```

```
                        * std;
            Fw[fmid] := (Fw[f1] + Fw[f2])/2.0 + gauss(0)
                * std;
            stdmid := std*ratio;
            subdivide(f1,fmid,stdmid);
            subdivide(fmid,f2,stdmid);
        end;
    end;

begin

    while seed <> 0 do
    begin
        CloseGraph;
        write('Enter seed (0 to quit): ');
        Readln(seed);
        GraphDriver := 4;
        GraphMode := EGAHi;
        InitGraph(graphDriver,GraphMode,'');
        SetLineStyle(0,$FFFF,1);
        SetColor(15);
        Fh[0] := gauss(seed) * scale;
        Fh[2048] := gauss(0) * scale;
        Fw[0] := gauss(seed) * scale;
        Fw[2048] := gauss(0) * scale;
        ratio := Exp(-0.693147*h);
        std := scale*ratio;
        subdivide(0,2048,std);
        for i:=0 to 2047 do
            Line(Round(Fw[i])+320,175-Round(Fh[i]),
                Round(Fw[i+1])+320,175-Round(Fh[i+1]));
        SetLineStyle(3,$0F0F,1);
        Line(80,175,590,175);
        Line(320,0,320,349);
        SetLineStyle(0,$FFFF,1);
        ch1 := ReadKey;
    end;
end.
```

Figure 18-4. Typical Two-Dimensional Brownian Motion

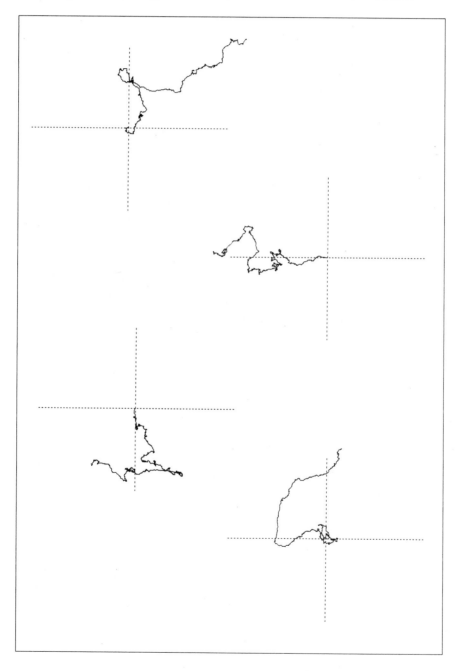

19

Fractal Landscapes

Some of the most interesting pictures created through the use of fractals are those which create extremely realistic landscape scenes. In Mandelbrot's book, *The Fractal Geometry of Nature*, there are scenes of rugged mountains and of a planet rising above a desolate landscape which were created by Richard Voss at the Thomas J. Watson Research Laboratory. These scenes have become almost classics in the art of fractal scene generation. We still look at these pictures in breathless amazement at their degree of realism. Other scenes including mountains, lakes, and oceans, have been created by Gary Mastin, Peter Watterberg, and John Mareda at Sandia National Laboratories.

Often the creators of these scenes are reluctant to reveal the details of the computer code that they use to produce them, so that we can only guess as to the exact details of how they were created. In general, the technique seems to be that of transforming Gaussian noise to the frequency domain, passing it through a 1/f filter, and then transforming the result back to the time domain. Using the result to generate three-dimensional shapes and then illuminating the shapes from a fictitious light source, using a ray-tracing program, appears to be the method to produce delightful landscapes. We take note, however, that a number of adjustments need to be made to the raw data before a lifelike scene appears. Since we don't have the code, we don't know exactly how much "tweaking" was done before these expressions of mathematics began to approach art.

The technique above can require an hour or more of processing time on a VAX computer; obviously it is not well suited to the capabilities of our PCs. In fact, attempts to adapt fractal scene generation to motion pictures encountered prohibitive costs in computer time (even for the movies, which are not noted for economizing), so that more efficient techniques were developed. Fournier, Fussell, and Carpenter developed the midpoint displacement method as an efficient and inexpensive method of generating fractal landscapes. Mandelbrot points out that this technique is not as mathematically sound as the one he was using, and consequently the landscapes are not as realistic, but for all practical purpose they are perfectly acceptable. Moreover, it is our good fortune that an adaptation of this technique can produce interesting scenes on a personal computer.

Midpoint Displacement Technique

Figure 19-1(a) shows the first step in the midpoint displacement process. We start with a triangle. We then go to the midpoint of each side of the triangle and displace it along the line and at right angles to the line. The amount of each displacement is determined by applying a Gaussian random multiplier to a proportion of the line length. We next connect each displaced midpoint to the two nearest apexes of the triangle. We then connect together each pair of displaced midpoints. Finally, we throw away the original sides of the triangle. (In the figure, we show the original triangle sides as wide lines, and the new lines as thin lines.) For clarity, we didn't throw away the original triangle lines in the figure; you can do that in your imagination.

The result of this process is that we have replaced the original triangle with four new triangles. We then apply the same process to each of the four new triangles, generating four more triangles from each, so that we then have sixteen triangles. This step is shown in Figure 19-1(c), although the sixteen new triangles are a little hard to pick out.

What happens when we have two triangles that have a common side? (Even if we start with a single triangle, this situation arises after the first set of displacements is performed.) If we are creating scenes on a large scale, using a mainframe computer, there is no problem in saving the coordinates of each new side as it is generated, so that once we have performed the midpoint displacement for a side, that same midpoint displacement is used for each triangle in which the side occurs.

For our smaller home computers, we prefer recursive techniques that are most efficient when we calculate the coordinates of one new triangle at the smallest level, fill it with some color, and then forget about it completely. Figure 19-1(b) shows what might happen when we do this. In this particular case, it happens that the displacement of the common side for the first triangle is toward the inside of the first triangle, and the displacement for the same side in the second triangle is toward the inside of the second triangle. If we assume that this is at the lowest level, and fill the resulting triangles with color, it is evident that there is a gap that is not filled, and will leave an objectionable unfilled space in the resulting picture.

There are two ways of approaching this problem. Michael Batty, at the University of Wales Institute of Science and Technology, has been applying this technique to BASIC programs for use with small computers. At each level of the recursion process except the lowest one, he creates a smaller triangle than the original and paints it with a color. Hopefully, this triangle is small enough so that it will not mask the desired irregularity of surfaces that are created by the midpoint displacement process, but large enough so that it will color in those regions where gaps might occur at the next downward level of the process.

Figure 19-1. Midpoint Displacement of Triangle Sides

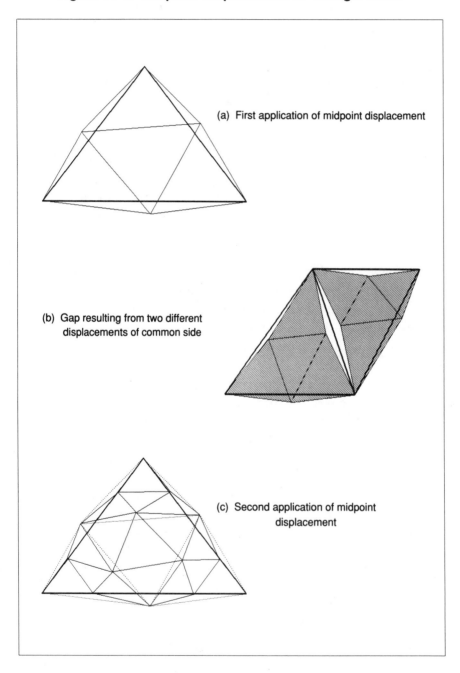

(a) First application of midpoint displacement

(b) Gap resulting from two different displacements of common side

(c) Second application of midpoint displacement

Another interesting approach is this. We use the coordinates of the undisplaced midpoint of the line that we are working on to generate a unique number. This number is then used as a seed for the random number generator of the computer, which then generates the random numbers for displacement along and perpendicular to the line. When the same line occurs in another triangle, since its undisplaced midpoint still has the same coordinates, the seed that is generated for the random number generator will be the same and the displacement will be the same, so that there is no possibility of a gap occurring.

Immediately the question arises, "Are these really random displacements?" The answer is, "Of course not!" But, then, what are they? They are certainly more random than displacements that might be generated by any arbitrary technique, but they are also much less random than if they were generated by truly random numbers that were created without the generator being reset for each line.

Interestingly enough, the resulting boundaries for this process are much rougher than those achieved by Batty's process, so that using the same basic generator data for the two processes does not result in pictures that are equally pleasing and seem equally realistic.

Oak Creek Canyon

Figure 19-2 lists a program for generating a western scene that is supposed to be reminiscent of Oak Creek Canyon near Sedona, Arizona. Before going into the program in detail, we need to consider what percentage of the picture realism is contributed by fractals, and what part is artistic endeavor by the programmer. To help get a handle on this, take a look at Figure 19-3, which is a line drawing showing the triangles and quadralaterals specified by data input to the program. This is the artist's contribution to the basic format of the picture. In addition, the parameters that are used with the function midpoint to specify the displacement of the midpoint of each line of each triangle have a lot to do with how rugged the landscape looks and its overall shape. Now look at the final picture as shown in Plate 29 (see color section) to see the result of combining artistic feeling with fractal scene generation.

The program to paint this scene uses the technique of specifying a unique seed for the random number generator at each midpoint generation so as to avoid objectionable gaps. First, let's look at the function *generate*. First, this function determines the length in x and y for each line making up the input triangle, and then calls midpoint to determine the x and y displacements of the midpoint for that line. The coordinates of the displaced midpoint are stored. If we are at the lowest level (level 0), the function next calls *plot_triangle* to determine the fill color and fill each of the four new triangles. If the lowest level has not been reached, the function calls node, which recursively calls *generate* to generate four new triangles for each of the four triangles that was just generated.

Take a quick look at the function *node* and you will see that it just returns if the level is zero and otherwise calls *generate* for each of the four triangles created by the current run of generate. Note that the way the program is set up, it should not ever be possible to enter node with the level set to zero. However, just in case, the first if statement provides added insurance that the program can never get into an endless loop.

For convenience, there is also a function *gen_quad*, which merely runs generate for two triangles which make up a quadralateral. Since many of the shapes in this scene are of a quadralateral nature, this function makes it easier to define them in the data.

Next, we'll look at the function *midpoint* in a little more detail, This function begins by selecting a random number which represents the displacement of the midpoint along the line whose x distance is stored in x, and whose y distance is stored in y. This distance is half the line length plus or minus a random value between zero and one-sixth times the line length. Next, the displacement at right angles to

the line is computed. It is the line length times a random number between 0.03 and 0.07, or between -0.03 and -0.07. Batty determines the
direction of the line and then applies trigonometric conversions to the
displacements along and at right angles to the line to determine the
displacements in the x and y directions. However, this is not really
necessary, since if you work out the trigonometry involved, you will
find that the sines and cosines of angles cancel out and the result is
those expressions given in the function for xz and yz.

Now let's take a look at the procedure *plot_triangle*. Its inputs are
the coordinates of the three apexes of a triangle and two color values
for the triangle. It also makes use of the global variable y_max
which is an altitude value that controls color selection.

Figure 19-2. Program to Generate Desert Scene

```
program sedona;

uses Graph,CRT,Fractal;

const
    color_set: array[0..15] of integer = (43,1,2,3,4,
        49,28,40,24,40,18,49,46,61,62,63);
    color_value: integer = 2;
    file_name: array[0..11] of char = 'sedona00.pcx';

var
    Triangle: array[1..3] of PointType;
    GraphDriver,GraphMode,interim,i,j,row,col: integer;
    y_max,x,y,xz,yz,xp,yp: real;
    ch1: char;
    file_no: string[3];
procedure generate(x1: integer; y1: integer; x2: integer;
    y2: integer; x3: integer; y3: integer;level: integer;
    color1: integer; color2: integer); forward;

procedure midpoint(x: real; y: real);

    var
        r,w: real;
        seed: longint;

    begin
        seed := Round(350*y + x);
```

```
        RandSeed := seed;
        r := 0.33333 + Random/3.0;
        w := 0.015 + Random/50.0;
        if Random < 0.5 then
            w := -w;
        xz := r*x - (w+0.05)*y;
        yz := r*y + (w +0.05)*x;
    end;

procedure node(x1: integer; y1: integer; x2: integer;
    y2:integer; x3: integer; y3: integer; x4: integer; y4:
    integer; x5: integer; y5: integer; x6: integer; y6:
    integer; level: integer; color1: integer; color2: integer);

    var
        x_ret1, y_ret1, x_ret2, y_ret2, x_ret3, y_ret3:
            integer;

    begin
        if level <> 0 then
        begin
            generate (x1,y1,x6,y6,x4,y4,
                level-1,color1,color2);
            generate (x2,y2,x4,y4,
                x5,y5,level-1,color1,color2);
            generate (x3,y3,x5,y5,x6,y6,
                level-1,color1,color2);
            generate (x4,y4,x5,y5,x6,y6,
                level-1,color1,color2);
        end;
    end;

procedure plot_triangle(x1: integer; y1: integer; x2:
    integer; y2: integer; x3: integer; y3: integer; color1:
    integer; color2: integer);

    var
        color,temp: integer;
        zt,ytt: real;

    begin
        if y1 > y2 then
            ytt := y1
        else
            ytt := y2;
        if ytt < y3 then
            ytt := y3;
        zt := 1 - ((ytt+240)*(ytt+240))
            /((y_max+240)*(y_max+240));
        if Random <= zt then
            color := color1
```

```
        else
            color := color2;
        if ytt + 240 <  ((y_max + 240)/4) then
            color := color1;
        if ytt+240 > (0.98 * (y_max+240)) then
            color := color2;
        Triangle[1].x := x1 + 320;
        Triangle[1].y := 175 - (longint(93*y1) div 128);
        Triangle[2].x := x2 + 320;
        Triangle[2].y := 175 - (longint(93*y2) div 128);
        Triangle[3].x := x3 + 320;
        Triangle[3].y := 175 - (longint(93*y3) div 128);
        SetColor(color);
        SetFillStyle(1,color);
        FillPoly(3,Triangle);
    end;

procedure generate(x1: integer; y1: integer; x2: integer;
    y2: integer; x3: integer; y3: integer;level: integer;
    color1: integer; color2: integer);

    var
        x4,x5,x6,y4,y5,y6: integer;

    begin
        x := x2 - x1;
        y := y2 - y1;
        midpoint(x,y);
        x4 := x1 + Round(xz);
        y4 := y1 + Round(yz);
        x := x1 - x3;
        y := y1 - y3;
        midpoint(x,y);
        x6 := x3 + Round(xz);
        y6 := y3 + Round(yz);
        x := x3 - x2;
        y := y3 - y2;
        midpoint(x,y);
        x5 := x2 + Round(xz);
        y5 := y2 + Round(yz);
        if level = 0 then
        begin
            plot_triangle(x1,y1,x6,y6,x4,y4,color1,color2);
            plot_triangle(x2,y2,x4,y4,x5,y5,color1,color2);
            plot_triangle(x3,y3,x5,y5,x6,y6,color1,color2);
            plot_triangle(x4,y4,x5,y5,x6,y6,color1,color2);
        end
        else
            node(x1,y1,x2,y2,x3,y3,x4,y4,x5,y5,x6,y6,
                level,color1, color2);
    end;
```

```pascal
procedure gen_quad (x1: integer; y1: integer; x2: integer;
    y2: integer; x3: integer; y3: integer; x4: integer; y4:
    integer;level: integer; color1: integer; color2: integer);

    begin
        generate(x1,y1,x2,y2,x3,y3,level,color1,color2);
        generate(x1,y1,x4,y4,x3,y3,level,color1,color2);
    end;

procedure cactus (x1: integer; y1: integer; scale: integer;
    level: integer; color1: integer; color2: integer);

    begin
        gen_quad(x1,y1,x1,y1+21*scale,x1+Round(1.6*scale),
            y1+22*scale,x1+ Round(1.6*scale),
            y1,level,color1,color2);
      gen_quad(x1+Round(1.4*scale),y1,x1+Round(1.4*scale),
            y1+22*scale,x1+3*scale,y1+21*scale,x1+3*scale,
            y1,level,color1,color2);
        gen_quad(x1,y1+9*scale,x1+7*scale,y1+9*scale,
            x1+7*scale,y1+12*scale, x1,y1+12*scale,0,
            color1,color2);
        gen_quad(x1,y1+9*scale,x1+6*scale,y1+9*scale,
            x1+7*scale,y1+12*scale, x1,y1+12*scale,level,
            color1,color2);
        gen_quad(x1+7*scale,y1+9*scale,x1+7*scale,
            y1+16*scale, x1+Round(8.5*scale),y1+17*scale,
            x1+Round(8.5*scale),y1+9*scale, level,color1,
            color2);
        gen_quad(x1+Round(8.4*scale),y1+9*scale,x1+
            Round(8.4*scale),y1+16*scale, x1+10*scale,y1+
            17*scale,x1+10*scale,y1+10*scale,
            level,color1,color2);
        gen_quad(x1,y1+7*scale,x1-6*scale,y1+7*scale,
            x1-6*scale,y1+10*scale, x1,y1+10*scale,0,
            color1,color2);
        gen_quad(x1,y1+7*scale,x1-6*scale,y1+7*scale,
            x1-6*scale,y1+10*scale, x1,y1+10*scale,level,
            color1,color2);
        gen_quad(x1-7*scale,y1+8*scale,x1-7*scale,
            y1+12*scale,x1-Round(5.4*scale), y1+13*scale,
            x1-Round(5.4*scale),y1+7*scale,level,color1,
            color2);
        gen_quad(x1-Round(5.6*scale),y1+7*scale,
            x1-Round(5.6*scale),y1+13*scale,x1-4*scale,
            y1+12*scale,x1-4*scale,y1+7*scale,level,
            color1,color2);
    end;

begin
    DirectVideo := false;
```

```
    GraphDriver := 4;
    GraphMode := EGAHi;
    InitGraph(graphDriver,GraphMode,'');
    for i:=0 to 15 do
        setEGApalette(i,color_set[i]);
    y_max := 280;
    gen_quad (-330,-150,-160,0,0,0,120,-130,4,4,12);
    gen_quad (-90,-110,85,50,200,50,440,-180,4,4,12);
    gen_quad (-160,-10,-160,220,-120,220,-120,-10,4,4,12);
    gen_quad (-120,-10,-120,190,-80,200,-80,-10,4,4,12);
    gen_quad (-80,-15,-80,230,-50,235,-50,-15,4,4,12);
    gen_quad (-50,-10,-50,100,0,180,0,-10,4,4,12);
    gen_quad (80,45,100,180,104,200,104,45,4,4,12);
    gen_quad (100,45,100,200,128,210,128,45,4,4,12);
    gen_quad (125,50,125,215,152,220,152,50,4,4,12);
    gen_quad (150,50,150,160,200,140,200,50,4,4,12);
    y_max := -100;
    gen_quad (-330,-300,-330,-110,330,-100,330,-300,4,6,14);
    y_max := 0;
    cactus(-110,-130,3,4,2,10);
    cactus(-200,-120,2,4,2,10);
    cactus(0,-160,4,4,2,10);
    cactus(210,-200,6,4,2,10);
    file_no := save_screen(0,0,640,350,file_name);
    ch1 := ReadKey;
end.
```

The procedure begins by selecting the y value of the highest apex of the triangle to use as the test criteria. It then creates a test variable, zt, according to the formula:

$$zt = (1 - ytt^2/y\_max^2) \qquad \text{(Equation 19-1)}$$

where y_max is the control altitude and ytt is the altitude of the highest apex of the triangle. Note, however, that both y's are in terms of system coordinates (which can take on any value between -240 and +240, so that in the procedure, 240 is added to every y value to make it a positive altitude). Once the value of zt is determined, the function selects a random number between zero and y_max and compares it with the test value zt. If the random number is less than or equal to zt, the first color is selected; otherwise, the second color is selected. Finally, if the altitude of ytt is below a selected limit, the color selection is overridden and the first color is selected; and if the altitude of ytt is above a selected limit, the color selection is overridden and the second color selected. The function then sets the x and y values of an array, which is then used with the *FillPoly* procedure to fill the triangle with the selected color.

Figure 19-3. Input Data for Oak Creek Canyon Scene

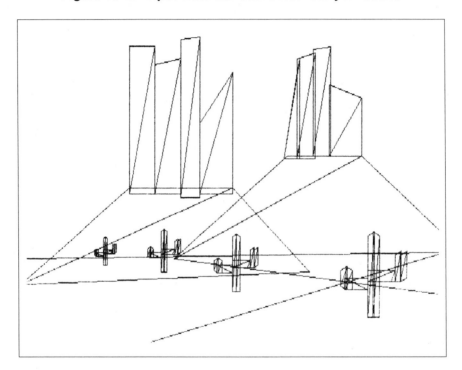

The *plot_triangle* function is where you can do a lot of experimentation to modify your scenes and provide them with additional color variations that are more complex than the ones used in the scenes depicted in this book. There is no reason why you need to be restricted to two colors; you can have several variations for different altitude levels. You don't need to have altitude be the criteria for color selection; you can select any criteria that you desire. You can also vary the controlling altitude and/or the overriding limits. Finally, there is the function *cactus*. This function has as inputs an x and y coordinate, a scale factor, a level, and two colors. It includes the necessary *gen_quad* function calls to create a picture of a cactus located at the x

and *y* coordinates, in a size determined by scale and with the specified level and colors.

With these preliminaries out of the way, we are ready to look at the main program. It first runs the *gen_quad* procedure several times to create the red rock cliffs. It then runs *gen_quad* once to paint the yellow and brown desert floor. It then runs cactus four times to create four cactii at different locations and of different sizes.

Pike's Peak at Sunrise

Figure 19-4 lists a program for generating a scene of Pike's Peak, near Colorado Springs, Colorado, at sunrise. Figure 19-5 is a line drawing showing the triangles specified by data input, which are the artist's contribution to the program. The final picture appears in Plate 30 (see color section). The program to paint this scene uses the technique of reseeding the random number generator for each line, prior to computing the midpoint displacement. The mountains are all created by using *generate* to fill triangles. Two blue triangles, which comprise a band across the screen, are filled with deep blue to depict the lake. The *generate* function is then used to create the foreground.

Figure 19-4. Program to Generate Pike's Peak Scene

```
program pikespk;

uses Graph,CRT,Fractal;

const
    color_set: array[0..15] of integer = (27,1,2,11,10,10,
        34,31,47,50,18,2,6,27,62,63);
    color_value: integer = 2;
    file_name: array[0..11] of char = 'pikspk00.pcx';

var
    Triangle: array[1..3] of PointType;
    GraphDriver,GraphMode,interim,i,j,row,col: integer;
    y_max,x,y,xz,yz,xp,yp: real;
    ch1: char;
```

```
    file_no: string[3];
 procedure generate(x1: integer; y1: integer; x2: integer;
    y2: integer; x3: integer; y3: integer;level: integer;
    color1: integer; color2: integer); forward;


procedure midpoint(x: real; y: real);

    var
        r,w: real;
        seed: longint;

    begin
        seed := Round(350*y + x);
        RandSeed := seed;
        r := 0.33333 + Random/3.0;
        w := 0.015 + Random/50.0;
        if Random < 0.5 then
            w := -w;
        xz := r*x - (w+0.05)*y;
        yz := r*y + (w +0.05)*x;
    end;

procedure node(x1: integer; y1: integer; x2: integer;
    y2:integer; x3: integer; y3: integer; x4: integer; y4:
    integer; x5: integer; y5: integer; x6: integer; y6:
    integer; level: integer; color1: integer; color2: integer);

    var
        x_ret1, y_ret1, x_ret2, y_ret2, x_ret3, y_ret3: integer;

    begin
        if level <> 0 then
        begin
            generate (x1,y1,x6,y6,x4,y4,level-1,
                color1,color2);
            generate (x2,y2,x4,y4,x5,y5,level-1,
                color1,color2);
            generate (x3,y3,x5,y5,x6,y6,level-1,
                color1,color2);
            generate (x4,y4,x5,y5,x6,y6,level-1,
                color1,color2);
        end;
    end;

procedure plot_triangle(x1: integer; y1: integer; x2:
    integer; y2: integer; x3: integer; y3: integer; color1:
    integer; color2: integer);

    var
        color,temp: integer;
```

```
        zt,ytt: real;

    begin
        if y1 > y2 then
            ytt := y1
        else
            ytt := y2;
        if ytt < y3 then
            ytt := y3;
        zt := 1 - ((ytt+240)*(ytt+240))/
            ((y_max+240)*(y_max+240));
        if Random <= zt then
            color := color1
        else
            color := color2;
        if ytt + 240 <  ((y_max + 240)/4) then
            color := color1;
        if ytt+240 > (0.98 * (y_max+240)) then
            color := color2;
        Triangle[1].x := x1 + 320;
        Triangle[1].y := 175 - (longint(93*y1) div 128);
        Triangle[2].x := x2 + 320;
        Triangle[2].y := 175 - (longint(93*y2) div 128);
        Triangle[3].x := x3 + 320;
        Triangle[3].y := 175 - (longint(93*y3) div 128);
        SetColor(color);          SetFillStyle(1,color);
        FillPoly(3,Triangle);
    end;

procedure generate(x1: integer; y1: integer; x2: integer;
    y2: integer; x3: integer; y3: integer;level: integer;
    color1: integer; color2: integer);

    var
        x4,x5,x6,y4,y5,y6: integer;

    begin
        x := x2 - x1;
        y := y2 - y1;
        midpoint(x,y);
        x4 := x1 + Round(xz);
        y4 := y1 + Round(yz);
        x := x1 - x3;
        y := y1 - y3;
        midpoint(x,y);
        x6 := x3 + Round(xz);
        y6 := y3 + Round(yz);
        x := x3 - x2;
        y := y3 - y2;
        midpoint(x,y);
        x5 := x2 + Round(xz);
```

```
        y5 := y2 + Round(yz);
        if level = 0 then
        begin
            plot_triangle(x1,y1,x6,y6,x4,y4,color1,color2);
            plot_triangle(x2,y2,x4,y4,x5,y5,color1,color2);
            plot_triangle(x3,y3,x5,y5,x6,y6,color1,color2);
            plot_triangle(x4,y4,x5,y5,x6,y6,color1,color2);
        end
        else
            node(x1,y1,x2,y2,x3,y3,x4,y4,x5,y5,x6,y6,level,
                color1,color2);
    end;
begin
    GraphDriver := 4;
    GraphMode := EGAHi;
    InitGraph(graphDriver,GraphMode,'');
    SetLineStyle(0,$FFFF,1);
    SetColor(15);
    for i:=0 to 15 do
        setEGApalette(i,color_set[i]);
    y_max := 180;
    generate(-220,-240,120,100,500,-40,6,3,7);
    y_max := 160;
    generate(-780,-200,30,130,480,-150,6,4,8);
    y_max := 180;
    generate(-480,0,-240,60,0,-60,5,5,9);
    generate(-100,-260,240,40,500,-180,5,6,10);
    Triangle[1].x := 0;
    Triangle[1].y := 320;
    Triangle[2].x := 0;
    Triangle[2].y := 225;
    Triangle[3].x := 639;
    Triangle[3].y := 225;
    SetColor(1);
    SetFillStyle(1,1);
    FillPoly(3,Triangle);
    Triangle[1].x := 639;
    Triangle[1].y := 225;
    Triangle[2].x := 639;
    Triangle[2].y := 320;
    Triangle[3].x := 0;
    Triangle[3].y := 320;
    FillPoly(3,Triangle);
    y_max := -100;
    generate(-770,-300,-250,-110,600,-300,5,11,12);
    generate(-550,-280,-60,-140,400,-300,5,11,12);
    generate(-220,-280,80,-130,340,-300,4,11,12);
    generate(-200,-280,230,-120,580,-300,4,11,12);
    file_no := save_screen(0,0,639,350,file_name);
    ch1 := ReadKey;
end.
```

Figure 19-5. Input Data for Pike's Peak Scene

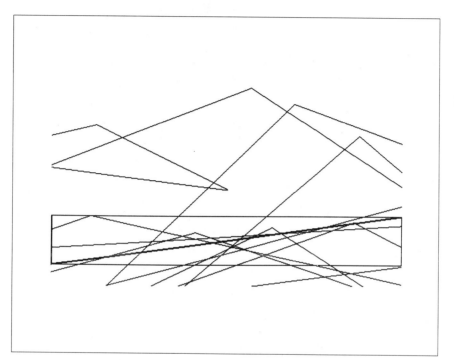

Earth Viewed from the Moon

Figure 19-6 lists a program for generating a scene of the earth viewed from the moon. Figure 19-7 is a line drawing showing the objects specified by data input, which are the artist's contribution to the program. The final picture appears in Plate 31 (see color section). Several new techniques are used in this program. First, note that the data for the planetary surface is referenced to an arbitrary zero reference point and that the actual call of *generate* modifies this reference to be the center of the planetary circle. The midpoint function is similar to the one used in the previous program but it uses different limits.

The *plot_triangle* function is the same as that used by the previous program, except that the limits for overriding the random color selection have been changed and a new type parameter has been added. For a type 0 selection, the same technique for selecting colors based on altitude is used, but this technique is not appropriate for the planetary surface, where altitude is not involved, so for painting this surface, a color selection based strictly upon a random number selection is used.

If we were not using Turbo Pascal's graphics procedure to fill the triangles which comprise the planet's land mass, we would write a new function for filling triangles which would clip off anything that tried to go outside the circle which is the planet. There is no easy way to do this with the built-in filling function, so we compromised by setting the parameters for creating the land mass so that the resulting land never went outside the planetary bounds.

This program also makes use of a new function, *ranFillOval*. We won't describe this function in detail. It is sufficient to say that it is like the function for filling an oval that is described in my book *Graphics Programming in C* except that it is set up so that when each point within the oval is determined, a random decision is made as to whether or not to plot this point. The probabilities are such that almost no points are plotted on the left side of the circle, and the probability of plotting points increases as one proceeds to the right of the circle.

After the planet is painted, this function is run with the same planetary circle defined and the background color as the designated color. It then blanks out some of the planetary circle to give the appearance of shading to indicate curvature of the sphere. The main program first sets the color palette, and then randomly plots 2000 white points to represent stars in the night sky. It then fills in a circle for the planet. Next it executes a *for* loop to paint the planetary surface on the circle. It then makes use of *ranFillOval* to provide the shading to give the circle the appearance of a sphere. Next, the *generate* procedure is used to create the moon's surface. Finally, several ovals are drawn on the moon's surface to represent meteor craters.

Figure 19-6. Program to Generate Earth Viewed from the Moon

```
program planet;

uses Graph,CRT,Fractal;

const
    color_set: array[0..15] of integer = (0,57,2,20,8,49,60,
        40,24,40,0,49,60,61,62,63);
    file_name: array[0..11] of char = 'planet00.pcx';
    xa: array[0..32] of integer = (-82,-80,-90,-70,-50,
        -30,-25,25,40,42,20,35,40,50,60,60,-28,70,
        -25,70,108,81,60,45,48,96,45,38,-8,0,-20,-28,55);
    xb: array[0..32] of integer = (-70,-70,-80,-50,-30,
        25,-25,40,65,65, 40,38,40,60,60,70,-28,90,
        -20,105,92,70,56,48,54,100,38,46,12,14,14,8,62);
    xc: array[0..32] of integer = (-70,-70,-80,-50,-50,
        20,30,40,40,58,40,37,50,50,70,75,20,90,
        -40,95,83,70,45,60,60,54,106,65,12,40,14,-8,95);
    xd: array[0..32] of integer = (-90,-80,-90,-70,-30,
        20,20,25,50,50,20,40,50,60,70,75,20,70,
        -40,90,81,108,45,56,96,45,100,106,8,44,0,-30,55);
    ya: array[0..32] of integer = (52,52,76,80,76,
        38,10,80,90,55,50,3,60,60,52,55,80,115,38,109,
        76,80,-130,-124,-90,-70,-60,-50,0,-10,10,90,-100);
    yb: array[0..32] of integer = (52,52,76,80,80,
        30,30,80,70,70,40,5,8,52,38,38,20,120, 78,104,
        58,95,-124,-90,-65,-60,-50,-25,0,-10,10,80,-100);
    yc: array[0..32] of integer = (60,80,80,55,56,38,10,
        90,80,70,3,-5,60,20,38,40,20,106, 78,76,60,110,
        -124,-100,-100,-65,-50,-25,-18,-30,-18,75,-50);
    yd: array[0..32] of integer= (60,80,77,55,38,30,30,
        90,60,60,3,-4,20,27,38,40,74,109,43,76,80,76,
        -124,-124,-70,-60,-50, -18,-30,-18,85,-50);
    color_value: integer = 2;
    level1: integer = 4;

var
    Triangle: array[1..3] of PointType;
    GraphDriver,GraphMode,interim,i,j,row,col,c_type:
        integer;
    y_max,x,y,xz,yz,xp,yp: real;
    ch1: char;
    file_no: string[3];
    x_center, y_center, radius: longint;

procedure generate(x1: integer; y1: integer; x2: integer;
    y2: integer; x3: integer; y3: integer;level: integer;
    color1: integer; color2: integer); forward;
```

```
procedure ranFillOval(x: integer; y:integer; b: integer;
    color: integer; aspect: real);
    var
        col,row,dummy,mask,end_x,end_y,kx,radius: integer;
        a,aspect_square: real;
        a_square,b_square,b_test,offset: longint;
    begin
        a := b/aspect;
        a_square := Round(a*a);
        radius := b;
        b := (93*b) div 128;
        b_square := b*b;
        x := x + 320;
        y := 175 - (93*y) div 128;
        end_x := x + Round(a);
        end_y := y + Round(b);
        for col:=x - Round(a) to end_x do
        begin
            b_test := b_square - (b_square*(col-x)*(col-x))
                div a_square;
            mask := $80 shr (col mod 8);
            Port[$3CE] := 8;
            Port[$3CF] := mask;
            Port[$3C4] := 2;
            Port[$3C5] := $0F;
            for row:=y-b to end_y do
            begin
                kx := Random(25205 div radius);
                if ((row-y)*(row-y) <= b_test) and (kx <
                    (col-x+20)) then
                begin
                    offset := row*80 + (col div 8);
                    dummy := byte(Mem[$A000:offset]);
                    Mem[$A000:offset] := 0;
                    Port[$3C4] := 2;
                    Port[$3C5] := color;
                    Mem[$A000:offset] := $FF;
                    Port[$3C4] := 2;
                    Port[$3C5] := $0F;
                end;
            end;
        end;
        Port[$3CE] := 3;
        Port[$3CF] := 0;
        Port[$3CE] := 8;
        Port[$3CF] := $FF;
    end;

procedure midpoint(x: real; y: real);

    var
```

```
        r,w: real;
        seed: longint;

    begin
        seed := Round(350*y + x);
        RandSeed := seed;
        r := 0.33333 + Random/3.0;
        w := 0.015 + Random/50.0;
        if Random < 0.5 then
            w := -w;
        xz := r*x - (w+0.05)*y;
        yz := r*y + (w +0.05)*x;
    end;

procedure node(x1: integer; y1: integer; x2: integer;
    y2:integer; x3: integer; y3: integer; x4: integer; y4:
    integer; x5: integer; y5: integer; x6: integer; y6:
    integer; level: integer; color1: integer; color2: integer);

    var
        x_ret1, y_ret1, x_ret2, y_ret2, x_ret3, y_ret3:
            integer;

    begin
        if level <> 0 then
        begin
            generate (x1,y1,x6,y6,x4,y4,level-1,
                color1,color2);
            generate (x2,y2,x4,y4,x5,y5,level-1,
                color1,color2);
            generate (x3,y3,x5,y5,x6,y6,level-1,
                color1,color2);
            generate (x4,y4,x5,y5,x6,y6,level-1,
                color1,color2);
        end;
    end;

procedure plot_triangle(x1: integer; y1: integer; x2:
    integer; y2: integer; x3: integer; y3: integer; color1:
    integer; color2: integer);

    var
        color,temp: integer;
        zt,ytt: real;

    begin
        if y1 > y2 then
            ytt := y1
        else
            ytt := y2;
        if ytt < y3 then
```

```
        ytt := y3;
    zt := 1 - ((ytt+240)*(ytt+240))/
        ((y_max+240)*(y_max+240));
    if c_type = 0 then
    begin
        if Random <= zt then
            color := color1
        else
            color := color2;
        if ytt + 240 <  ((y_max + 240)/4) then
            color := color1;
        if ytt+240 > (0.98 * (y_max+240)) then
            color := color2;
    end
    else
    begin
        if Random < 0.75 then
            color := color1
        else
            color := color2;
    end;
    Triangle[1].x := x1 + 320;
    Triangle[1].y := 175 - (longint(93*y1) div 128);
    Triangle[2].x := x2 + 320;
    Triangle[2].y := 175 - (longint(93*y2) div 128);
    Triangle[3].x := x3 + 320;
    Triangle[3].y := 175 - (longint(93*y3) div 128);
    SetColor(color);
    SetFillStyle(1,color);
    FillPoly(3,Triangle);
end;

procedure generate(x1: integer; y1: integer; x2: integer;
    y2: integer; x3: integer; y3: integer;level: integer;
    color1: integer; color2: integer);

    var
        x4,x5,x6,y4,y5,y6: integer;

    begin
        x := x2 - x1;
        y := y2 - y1;
        midpoint(x,y);
        x4 := x1 + Round(xz);
        y4 := y1 + Round(yz);
        x := x1 - x3;
        y := y1 - y3;
        midpoint(x,y);
        x6 := x3 + Round(xz);
        y6 := y3 + Round(yz);
        x := x3 - x2;
```

```
        y := y3 - y2;
        midpoint(x,y);
        x5 := x2 + Round(xz);
        y5 := y2 + Round(yz);
        if level = 0 then
        begin
            plot_triangle(x1,y1,x6,y6,x4,y4,color1,color2);
            plot_triangle(x2,y2,x4,y4,x5,y5,color1,color2);
            plot_triangle(x3,y3,x5,y5,x6,y6,color1,color2);
            plot_triangle(x4,y4,x5,y5,x6,y6,color1,color2);
        end
        else
        node(x1,y1,x2,y2,x3,y3,x4,y4,x5,y5,x6,y6,
            level,color1,color2);
    end;

procedure gen_quad (x1: integer; y1: integer; x2: integer;
    y2: integer; x3: integer; y3: integer; x4: integer; y4:
    integer;level: integer; color1: integer; color2: integer);

    begin
        generate(x1,y1,x2,y2,x3,y3,level,color1,color2);
        generate(x1,y1,x4,y4,x3,y3,level,color1,color2);
    end;


begin
    DirectVideo := false;
    GraphDriver := 4;
    GraphMode := EGAHi;
    InitGraph(graphDriver,GraphMode,'');
    SetLineStyle(0,$FFFF,1);
    SetColor(15);
    for i:=0 to 15 do
        setEGApalette(i,color_set[i]);
    x_center := -100;
    y_center := 0;
    radius := 150;
    for i:=0 to 2000 do
    begin
        row := Random(349);
        col := Random(639);
        PutPixel(col,row,15);
    end;
    SetFillStyle(1,1);
    SetColor(1);
    FillEllipse(220,175,152,106);
    y_max := 280;
    c_type := 1;
    for i:=0 to 32 do
        gen_quad(xa[i]+x_center,ya[i]+y_center,xb[i]+
```

```
            x_center,yb[i]+y_center,xc[i]+x_center,yc[i]+
            y_center,xd[i]+x_center,yd[i]+y_center,
            level1,2,3);
    ranFillOval(-100,0,154,0,1.0);
    c_type := 0;
    y_max := -80;
    generate(-470,-300,-250,-100,300,-300,4,14,6);
    generate(-350,-280,-60,-120,300,-300,4,14,6);
    generate(-220,-280,80,-110,340,-300,4,14,6);
    generate(-200,-280,230,-100,580,-300,4,14,6);
    SetFillStyle(1,6);
    SetColor(6);
    FillEllipse(140,320,20,7);
    FillEllipse(290,310,20,7);
    FillEllipse(360,334,48,10);
    FillEllipse(420,300,12,4);
    FillEllipse(520,313,22,9);
    FillEllipse(100,270,16,5);
    FillEllipse(600,284,16,5);
    file_no := save_screen(0,0,639,349,file_name);
    ch1 := ReadKey;
end.
```

Figure 19-7. Input Data for Moon Scene

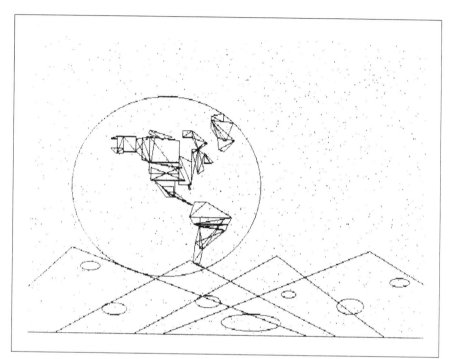

Lights of Albuquerque

This scene is supposed to be reminiscent of the lights of Albuquerque at night. It begins by distributing random points over the black sky background to represent stars, but with considerably less stars than would be seen from the surface of the moon. A full moon is then created with the *FillEllipse* procedure, and is then shaded with the *ranFillOval* procedure to give the appearance of roundness. Next, the midpoint displacement technique is used to create a mountain background in appropriate colors for a night scene. Finally, a mixture of dotted lines and random placement of points is used to attempt to represent the city lights with both suggestions of randomness and of a grid of streets. This shows how both fractal and non-fractal tech-

niques can be combined to create an interesting scene. This scene does
not appear in the color section, but it is well worth generating on your
PC. The listing of the program to generate this scene is given in
Figure 19-8.

Figure 19-8. Program to Create Lights of Albuquerque

```
program sandia;

uses Graph,CRT,Fractal;

const
    color_set: array[0..15] of integer =(0,1,2,3,8,49,33,40,
        24,40,0,49,60,61,62,63);
    color_value: integer = 2;
    file_name: array[0..11] of char = 'sandia00.pcx';

var
    Triangle: array[1..3] of PointType;
    GraphDriver,GraphMode,interim,i,j,row,col: integer;
    y_max,x,y,xz,yz,xp,yp: real;
    ch1: char;
    file_no: string[3];

procedure generate(x1: integer; y1: integer; x2: integer;
    y2: integer; x3: integer; y3: integer;level: integer;
    color1: integer;color2: integer); forward;

procedure ranFillOval(x: integer; y:integer; b: integer;
    color: integer; aspect: real);

    var
        col,row,dummy,mask,end_x,end_y,kx,radius: integer;
        a,aspect_square: real;
        a_square,b_square,b_test,offset: longint;

    begin
        a := b/aspect;
        a_square := Round(a*a);
        radius := b;
        b := (93*b) div 128;
        b_square := b*b;
        x := x + 320;
        y := 175 - (93*y) div 128;
        end_x := x + Round(a);
        end_y := y + Round(b);
        for col:=x - Round(a) to end_x do
        begin
```

```
                    b_test := b_square - (b_square*(col-x)*(col-x))
                        div a_square;
                    mask := $80 shr (col mod 8);
                    Port[$3CE] := 8;
                    Port[$3CF] := mask;
                    Port[$3C4] := 2;
                    Port[$3C5] := $0F;
                    for row:=y-b to end_y do
                    begin
                        kx := Random(25205 div radius);
                        if ((row-y)*(row-y) <= b_test) and (kx <
                            (col-x+20)) then
                        begin
                            offset := row*80 + (col div 8);
                            dummy := byte(Mem[$A000:offset]);
                            Mem[$A000:offset] := 0;
                            Port[$3C4] := 2;
                            Port[$3C5] := color;
                            Mem[$A000:offset] := $FF;
                            Port[$3C4] := 2;
                            Port[$3C5] := $0F;
                        end;
                    end;
                end;
                Port[$3CE] := 3;
                Port[$3CF] := 0;
                Port[$3CE] := 8;
                Port[$3CF] := $FF;
        end;

procedure midpoint(x: real; y: real);

        var
            r,w: real;
            seed: longint;

        begin
            seed := Round(350*y + x);
            RandSeed := seed;
            r := 0.33333 + Random/3.0;
            w := 0.015 + Random/50.0;
            if Random < 0.5 then
                w := -w;
            xz := r*x - (w+0.05)*y;
            yz := r*y + (w +0.05)*x;
        end;

procedure node(x1: integer; y1: integer; x2: integer;
    y2:integer; x3: integer; y3: integer; x4: integer; y4:
    integer; x5: integer; y5: integer; x6: integer; y6:
    integer; level: integer; color1: integer; color2: integer);
```

```
    var
        x_ret1, y_ret1, x_ret2, y_ret2, x_ret3, y_ret3:
            integer;

    begin
        if level <> 0 then
        begin
            generate (x1,y1,x6,y6,x4,y4,level-1,
                color1,color2);
            generate (x2,y2,x4,y4,x5,y5,level-1,
                color1,color2);
            generate (x3,y3,x5,y5,x6,y6,level-1,
                color1,color2);
            generate (x4,y4,x5,y5,x6,y6,level-1,
                color1,color2);
        end;
    end;

procedure plot_triangle(x1: integer; y1: integer; x2:
    integer; y2: integer; x3: integer; y3: integer; color1:
    integer; color2: integer);

    var
        color,temp: integer;
        zt,ytt: real;

    begin
        if y1 > y2 then
            ytt := y1
        else
            ytt := y2;
        if ytt < y3 then
            ytt := y3;
        zt := 1 - ((ytt+240)*(ytt+240))/
            ((y_max+240)*(y_max+240));
        if Random <= zt then
            color := color1
        else
            color := color2;
        if ytt + 240 < ((y_max + 240)/4) then
                color := color1;
        if ytt+240 > (0.98 * (y_max+240)) then
            color := color2;
        Triangle[1].x := x1 + 320;
        Triangle[1].y := 175 - (longint(93*y1) div 128);
        Triangle[2].x := x2 + 320;
        Triangle[2].y := 175 - (longint(93*y2) div 128);
        Triangle[3].x := x3 + 320;
        Triangle[3].y := 175 - (longint(93*y3) div 128);
        SetColor(color);
```

```
        SetFillStyle(1,color);
        FillPoly(3,Triangle);
    end;

procedure generate(x1: integer; y1: integer; x2: integer;
    y2: integer; x3: integer; y3: integer;level: integer;
    color1: integer; color2: integer);

    var
        x4,x5,x6,y4,y5,y6: integer;

    begin
         x := x2 - x1;
        y := y2 - y1;
        midpoint(x,y);
        x4 := x1 + Round(xz);
        y4 := y1 + Round(yz);
        x := x1 - x3;
        y := y1 - y3;
        midpoint(x,y);
        x6 := x3 + Round(xz);
        y6 := y3 + Round(yz);
        x := x3 - x2;
        y := y3 - y2;
        midpoint(x,y);
        x5 := x2 + Round(xz);
        y5 := y2 + Round(yz);
        if level = 0 then
        begin
            plot_triangle(x1,y1,x6,y6,x4,y4,color1,color2);
            plot_triangle(x2,y2,x4,y4,x5,y5,color1,color2);
            plot_triangle(x3,y3,x5,y5,x6,y6,color1,color2);
            plot_triangle(x4,y4,x5,y5,x6,y6,color1,color2);
        end
        else
            node(x1,y1,x2,y2,x3,y3,x4,y4,x5,y5,x6,y6,level,
                color1,color2);
    end;

procedure gen_quad (x1: integer; y1: integer; x2: integer;
    y2: integer; x3: integer; y3: integer; x4: integer; y4:
    integer;level: integer; color1: integer; color2: integer);

    begin
        generate(x1,y1,x2,y2,x3,y3,level,color1,color2);
        generate(x1,y1,x4,y4,x3,y3,level,color1,color2);
    end;


begin
    DirectVideo := false;
```

```
GraphDriver := 4;
GraphMode := EGAHi;
InitGraph(graphDriver,GraphMode,'');
SetLineStyle(0,$FFFF,1);
SetColor(15);
for i:=0 to 15 do
    setEGApalette(i,color_set[i]);
for i:=0 to 750 do
begin
    row := Random(349);
    col := Random(639);
    PutPixel(col,row,15);
end;
SetFillStyle(1,14);
SetColor(14);
FillEllipse(220,90,80,56);
ranFillOval(-100,117,80,0,1.0);
y_max := 160;
generate(-480,0,-260,140,0,-60,6,5,9);
gen_quad (-980,-200,40,110,210,120,420,-120,6, 4,8);
y_max := 180;
generate(-100,-260,280,110,500,-180,5,7,11);
Triangle[1].x := 0;
Triangle[1].y := 365;
Triangle[2].x := 0;
Triangle[2].y := 220;
Triangle[3].x := 639;
Triangle[3].y := 220;
SetColor(0);
SetFillStyle(1,0);
FillPoly(3,Triangle);
Triangle[1].x := 639;
Triangle[1].y := 220;
Triangle[2].x := 639;
Triangle[2].y := 365;
Triangle[3].x := 0;
Triangle[3].y := 365;
FillPoly(3,Triangle);
y_max := -100;
SetColor(14);
SetLineStyle(4,$0101,1);
Line(320,220,320,349);
Line(360,220,440,349);
Line(400,220,560,349);
Line(440,220,639,335);
Line(480,220,639,290);
Line(520,220,639,262);
Line(290,220,230,349);
Line(260,220,140,349);
Line(230,220,50,349);
Line(200,220,0,335);
```

```
        Line(170,220,0,305);
        Line(140,220,0,277);
        Line(0,255,639,255);
        Line(0,285,639,284);
        SetLineStyle(4,$0003,3);
        Line(0,316,639,313);
        Line(0,342,639,342);
        for i:=0 to 500 do
        begin
            row := random(80) + 220;
            col := random(640);
            PutPixel(col,row,14);
        end;
        for i:=0 to 75 do
        begin
            row := random(20) + 329;
            col := random(640);
            PutPixel(col,row,14);
            PutPixel(col,row,14);
            PutPixel(col-1,row,14);
            PutPixel(col+1,row,14);
            PutPixel(col,row-1,14);
            PutPixel(col,row+1,14);
        end;
        for i:=0 to 100 do
        begin
            row := random(100) + 249;
            col := random(640);
            PutPixel(col,row,14);
            PutPixel(col,row,14);
            PutPixel(col-1,row,14);
            PutPixel(col+1,row,14);
            PutPixel(col,row-1,14);
            PutPixel(col,row+1,14);
        end;
        file_no := save_screen(0,0,639,349,file_name);
        ch1 := ReadKey;
end.
```

Forest Scene

This scene is a demonstration of how two different kinds of fractals can be combined to create an interesting scene. The mountain background is created using the midpoint displacement technique in the same manner that is described for the several scenes that have just been described in this chapter. The forest is generated by using a variation of the trees program that was described in Chapter 10.

First, the program generates 20 random pairs of coordinates to serve as locations for the base of the trunks of the trees. These are then sorted on their *y* coordinates. The tree with the highest *y* coordinate needs to be drawn first because it is farthest back in the picture, so that trees with progressively lower y coordinates will overlap the higher ones. The parameters of the tree generating program have also been adapted so that the branch length and branch width are proportional to the y coordinate so that the trees in the background will be smaller than the trees in the foreground.

Finally, the other tree parameters have been given a basic set of parameters and then perturbed with a random amount so that the shape of each tree will be slightly different from all others. This scene does not appear in the color section. The program to generate this scene is listed in Figure 19-9.

Figure 19-9. Program to Generate Forest Scene

```
program forest;

uses CRT, Graph,Fractal;

const
    color_set: array[0..15] of integer = (43,33,2,34,16,35,
        20,48,48,45,58,2,6,27,62,63);
    ln_2: real = (0.6931471);
    Square: array[1..4] of PointType =    ((x: 50;   y: 100),
                                (x: 100;  y: 100),
                                (x: 100;  y: 100),
                                (x: 150;  y: 150));
    file_name: array[0..11] of char = 'forest00.pcx';

var

    GraphDriver,GraphMode,i,j,level,width,indx,jndx: integer;
    Triangle: array[1..3] of PointType;
    height,left_alpha,right_alpha,left_angle,right_angle,
        left_width_factor,left_height_factor,
        right_width_factor,right_height_factor,x,y,x1,y1,
        x2,y2,val1,val2: real;
    list: array[0..20,0..1] of real;
    y_max,xz,yz,xp,yp: real;
```

```
        file_no: string[3];
        ch: char;

procedure generate(x1: integer; y1: integer; x2: integer;
    y2: integer; x3: integer; y3: integer;level: integer;
    color1: integer; color2: integer); forward;


procedure midpoint(x: real; y: real);

    var
        r,w: real;
        seed: longint;

    begin
        seed := Round(350*y + x);
        RandSeed := seed;
        r := 0.33333 + Random/3.0;
        w := 0.015 + Random/50.0;
        if Random < 0.5 then
            w := -w;
        xz := r*x - (w+0.05)*y;
        yz := r*y + (w +0.05)*x;
    end;

procedure node(x1: integer; y1: integer; x2: integer;
    y2:integer; x3: integer; y3: integer; x4: integer; y4:
    integer; x5: integer; y5: integer; x6: integer; y6:
    integer; level: integer; color1: integer; color2: integer);
var
        x_ret1, y_ret1, x_ret2, y_ret2, x_ret3, y_ret3:
            integer;

    begin
        if level <> 0 then
        begin
            generate(x1,y1,x6,y6,x4,y4, level-1,color1,color2);
            generate (x2,y2,x4,y4,x5,y5,level-1,
                color1,color2);
            generate (x3,y3,x5,y5,x6,y6,level-1,
                color1,color2);
            generate (x4,y4,x5,y5,x6,y6,level-1,
                color1,color2);
        end;
    end;

procedure plot_triangle(x1: integer; y1: integer; x2:
    integer; y2: integer; x3: integer; y3: integer; color1:
    integer; color2: integer);

    var
```

```
        color,temp: integer;
        zt,ytt: real;

    begin
        if y1 > y2 then
            ytt := y1
        else
            ytt := y2;
        if ytt < y3 then
            ytt := y3;
        zt := 1 - ((ytt+240)*(ytt+240))/
            ((y_max+240)*(y_max+240));
        if Random <= zt then
            color := color1
        else
            color := color2;
        if ytt + 240 <  ((y_max + 240)/4) then
            color := color1;
        if ytt+240 > (0.98 * (y_max+240)) then
            color := color2;
        Triangle[1].x := x1 + 320;
        Triangle[1].y := 175 - (longint(93*y1) div 128);
        Triangle[2].x := x2 + 320;
        Triangle[2].y := 175 - (longint(93*y2) div 128);
        Triangle[3].x := x3 + 320;
        Triangle[3].y := 175 - (longint(93*y3) div 128);
        SetColor(color);
        SetFillStyle(1,color);
        FillPoly(3,Triangle);
    end;

procedure generate(x1: integer; y1: integer; x2: integer;
    y2: integer; x3: integer; y3: integer;level: integer;
    color1: integer; color2: integer);

    var
        x4,x5,x6,y4,y5,y6: integer;

    begin
        x := x2 - x1;
        y := y2 - y1;
        midpoint(x,y);
        x4 := x1 + Round(xz);
        y4 := y1 + Round(yz);
        x := x1 - x3;
        y := y1 - y3;
        midpoint(x,y);
        x6 := x3 + Round(xz);
        y6 := y3 + Round(yz);
        x := x3 - x2;
        y := y3 - y2;
```

```
        midpoint(x,y);
        x5 := x2 + Round(xz);
        y5 := y2 + Round(yz);
        if level = 0 then              begin
            plot_triangle(x1,y1,x6,y6,x4,y4,color1,color2);
            plot_triangle(x2,y2,x4,y4,x5,y5,color1,color2);
            plot_triangle(x3,y3,x5,y5,x6,y6,color1,color2);
            plot_triangle(x4,y4,x5,y5,x6,y6,color1,color2);
        end
        else
        node(x1,y1,x2,y2,x3,y3,x4,y4,x5,y5,x6,y6,level,
            color1,color2);
    end;

procedure t_generate(x: real; y: real; width: integer;
    height: real; angle: real; level: integer);

    var
        x1,y1: real;

    begin
        turtle_x := x;
        turtle_y := y;
        turtle_r := height;
        step;
        x1 := turtle_x;
        y1 := turtle_y;
        dec(level);
        if level < 3 then
        begin
            SetColor(10);
            SetFillStyle(1,10);
        end
        else
        begin
            SetColor(6);
            SetFillStyle(1,6);
        end;
        if Abs(x - x1) > Abs(y - y1) then
        begin
            Square[1].x := Round(x+320);
            Square[1].y := Round(175-y*0.729) + width div 2;
            Square[2].x := Round(x+320);
            Square[2].y := Round(175-y*0.729) - width div 2;
            Square[3].x := Round(x1+320);
            Square[3].y := Round(175-y1*0.729)-width div 2;
            Square[4].x := Round(x1+320);
            Square[4].y := Round(175-y1*0.729)+width div 2;
        end
        else
        begin
```

```
            Square[1].x := Round(x+320) - width div 2;
            Square[1].y := Round(175 - y*0.729);
            Square[2].x := Round(x+320) + width div 2;
            Square[2].y := Round(175 - y*0.729);
            Square[3].x := Round(x1+320) + width div 2;
            Square[3].y := Round(175 - y1*0.729);
            Square[4].x := Round(x1+320) - width div 2;
            Square[4].y := Round(175 - y1*0.729);
        end;
        FillPoly(4,Square);
        if level > 0 then
        begin
            turtle_theta := point(x,y,x1,y1);
            turn(left_angle);
            t_generate(x1,y1,Round(left_width_factor*width),
                left_height_factor*height,left_angle,level);
            turtle_theta := point(x,y,x1,y1);
            turn(-right_angle);
            t_generate(x1,y1,Round(right_width_factor*width),
                right_height_factor*height,right_angle,level);
        end;
    end;

procedure trees(x: real; y: real);

    begin
        height := Round((240 - y)/12);
        width := Round((240 - y)/48);
        left_alpha := 1.3 + Random;
        right_alpha := 1.3 + Random;
        left_angle := 20.0 + 5*Random;
        right_angle := 20.0 + 5*Random;
        level := 14;
        left_width_factor := exp((-1/left_alpha)*ln_2);
        left_height_factor := exp((-2/(3*left_alpha))*ln_2);
        right_width_factor := exp((-1/right_alpha)*ln_2);
        right_height_factor := exp((-2/(3*right_alpha))
            *ln_2);
        x1 := x;
        y1 := y + height;
        SetColor(6);
        SetFillStyle(1,6);
        if Abs(x - x1) > Abs(y - y1) then
        begin
            Square[1].x := Round(x+320);
            Square[1].y := Round(175-y*0.729) + width div 2;
            Square[2].x := Round(x+320);
            Square[2].y := Round(175-y*0.729) - width div 2;
            Square[3].x := Round(x1+320);
            Square[3].y := Round(175-y1*0.729)-width div 2;
            Square[4].x := Round(x1+320);
```

```
            Square[4].y := Round(175-y1*0.729)+width div 2;
        end
        else
        begin
            Square[1].x := Round(x+320) - width div 2;
            Square[1].y := Round(175 - y*0.729);
            Square[2].x := Round(x+320) + width div 2;
            Square[2].y := Round(175 - y*0.729);
            Square[3].x := Round(x1+320) + width div 2;
            Square[3].y := Round(175 - y1*0.729);
            Square[4].x := Round(x1+320) - width div 2;
            Square[4].y := Round(175 - y1*0.729);
        end;
        FillPoly(4,Square);
        turtle_theta := point(x,y,x1,y1);
        turn(left_angle);
        t_generate(x1,y1,Round(left_width_factor*width),
            left_height_factor*height,left_angle,level);
        turtle_theta := point(x,y,x1,y1);
        turn(-right_angle);
        t_generate(x1,y1,Round(right_width_factor*width),
            right_height_factor*height,right_angle,level);
    end;

begin
    DirectVideo := false;
    GraphDriver := 4;
    GraphMode := EGAHi;
    InitGraph(graphDriver,GraphMode,'');
    SetBkColor(9);
    ClearDevice;
    for i:=0 to 15 do
        setEGApalette(i,color_set[i]);
    y_max := 180;
    generate(-220,-240,220,125,500,-40,6,1,3);
    y_max := 160;
    generate(-880,-200,30,125,680,-150,6,4,5);
    y_max := 180;
    generate(-580,0,-240,100,150,-60,5,7,10);
    generate(-300,-260,200,40,500,-180,5,8,11);
    Square[1].x := 0;
    Square[1].y := 180;
    Square[2].x := 639;
    Square[2].y := 180;
    Square[3].x := 639;
    Square[3].y := 349;
    Square[4].x := 0;
    Square[4].y := 349;
    SetColor(2);
    SetFillStyle(1,2);
    FillPoly(4,Square);
```

```
    for i:= 0 to 20 do
    begin
        list[i,0] := 639*Random - 320;
        list[i,1] := 230*Random - 240;
    end;
    for indx:=0 to 19 do
    begin
        for jndx:= jndx to 20 do
        begin
            if list [jndx,1] <
                list [jndx,1] then
        val1 := list[jndx,0];
        val2 := list[jndx,1];
            list[jndx,0] := list[indx,0];
            list[jndx,1] := list[indx,1];
        list[indx,0] := val1;
        list[indx,1] := val2;
        end;
        end;
    end;
    for i:=20 downto 0 do
    begin
        x := list[i,0];
        y := list[i,1];
        trees(x,y);
    end;
    file_no := save_screen(0,0,639,350,file_name);
    ch := ReadKey;
end.
```

20

Iterated Function Systems

Michael Barnsley at the Georgia Institute of Technology calls it the "Chaos Game," or more formally, "Iterated Function Systems." A point is moved about the screen, randomly being translated, rotated, and scaled by one of several affine transformations. Ultimately, an image of the attractors is produced. We have already seen this technique at work in Chapter 4 where it generated the image of some strange attractors, in Chapter 14, where it produced the outlines of dragon curves, and in Chapter 16, where it generated three-dimensional dragons. Barnsley uses the technique to generate scenes in which a minimum of input data and a simple conversion program produce tremendous amounts of detail, representative of compression ratios of up to 10,000 compared to the original pixel data needed to create a similar scene.

Affine Transformations

The primary tool used in generating scenes with iterated function systems (IFS) is the affine transformation. As used here, this is a rotation, translation, and scaling of the coordinates of a point on the display screen (x,y) to a new position represented by (x_n, y_n). The transformation is performed as follows:

$$\begin{bmatrix} xn \\ yn \end{bmatrix} = w \begin{bmatrix} x \\ y \end{bmatrix} = \begin{bmatrix} a & b \\ c & d \end{bmatrix} \begin{bmatrix} x \\ y \end{bmatrix} + \begin{bmatrix} e \\ f \end{bmatrix}$$

Equation 20-1

$$= \begin{bmatrix} ax + by + e \\ cx + dy + f \end{bmatrix}$$

The parameters $a, b, c,$ and d perform a rotation, and their magnitudes result in the scaling. (For the whole system to work properly, the scaling must always result in a shrinkage of the distances between points; otherwise, repeated iterations will result in the function blowing up to infinity, which is something your computer won't like. The parameters e and f cause a linear translation of the point being operated upon. Given that you apply this transformation to a geometric figure, the figure will be translated to a new location, and rotated and shrunk to a new, smaller size.

The Deterministic Algorithm

Before we begin playing the "chaos game," we should note that there is a deterministic version of the method. When we use this method, we take each point on our display screen and apply to it each of the affine transformations that make up our IFS for a particular desired figure. The new points are then plotted, and then the same process is applied again, as many times as necessary to obtain a final result. Figure 20-1 lists the parameters required to generate a Sierpinski triangle and a fern leaf using the deterministic algorithm. (You probably thought that by the time we were through with Chapter 9 you had seen every possible way to generate a Sierpinski triangle, but you were wrong; two more methods are given in this chapter.) Figure 20-2 lists the program to generate figures using the deterministic algorithm.

At this point, we need to be aware of two significant problems with the deterministic program. First, we need to have a lot of memory available. Just to specify all of the points on our current screen requires 224K of memory locations, and a like number of locations are required for storing the transformed points as we generate them.

That is why a deterministic program in Barnsley's book, *Fractals Everywhere*, limits the resulting picture to a tiny size of 100 x 100 pixels. We get around the problem by making use of the two screens that are available for graphics with the EGA when it has its full memory complement. This makes the program a little slow, but still practical. We have to be careful in using this technique to specify which of the two graphics pages we are using for various operations.

We begin by writing our starting figure (a rectangle) out to screen 0. Then, we use two *for* loops to read each pixel from the screen in turn. If the pixel is not black (the background color) we apply each affine transformation to the point and plot all the new points generated on screen 1. Eight of these iterations are processed to generate the Sierpinski triangle; 32 are used for the fern leaf. For the second iteration, points are read from screen 1 and the transformed points are stored on screen 0. For each successive iteration, the screens are reversed. The program first generates the Sierpinski triangle. Figure 20-3 shows the original rectangle, the first iteration of the program, and two iterations further along. After about eight iterations, no further change in the program can be observed on the screen.

Figure 20-1. Parameters for Deterministic Generation of Sierpinski Triangle and Fern Leaf

Figure	a	b	c	d	e	f
Sierpinski Triangle	0.5	0.0	0.0	0.5	75.0	0.0
	0.5	0.0	0.0	0.5	0.0	150.0
	0.5	0.0	0.0	0.5	150.0	150.0
Fern Leaf	0.0	0.0	0.0	0.16	0.0	0.0
	0.85	0.04	-0.04	0.85	0.0	40.0
	0.2	-0.26	0.23	0.22	0.0	40.0
	-0.15	0.28	0.26	0.24	0.0	10.0

Figure 20-2. Program to Generate Figures
Using Deterministic Algorithm

```pascal
program ifsdet;

uses CRT,Graph,Fractal;

const

    a: array[0..3] of real = (0.5 ,0.5, 0.5,0);
    b: array[0..3] of real = (0,0,0,0);
    c: array[0..3] of real = (0,0,0,0);
    d: array[0..3] of real = (0.5,0.5,0.5,0);
    e: array[0..3] of real = (75,0,150,0);
    f: array[0..3] of real = (0,150,150,0);
    dis_page: integer = 1;
    back_page: integer = 0;
    iterations: integer = 8;
    start_col: integer = 0;
    set_end: integer = 3;

var
    GraphDriver,GraphMode,i,j,k,m,iter,color,displays,
        temp,index: integer;
    Xpoints, Ypoints: array[0..24] of real;
    initiator_x1,initiator_x2,initiator_y1,initiator_y2:
        array[0..9] of real;
    ch1: char;

begin
    GraphDriver := 4;
    GraphMode := EGAHi;
    InitGraph(graphDriver,GraphMode,'');
    for displays:=0 to 1 do
    begin
        SetVisualPage(0);
        SetActivePage(0);
        ClearDevice;
        for i:=2 to 297 do
        begin
            PutPixel(i+100,0,15);
            PutPixel(i+100,299,15);
            PutPixel(399,i,15);
            PutPixel(100,i,15);
        end;
        for index:=0 to iterations do
        begin
            SetActivePage(dis_page);
            SetVisualPage(dis_page);
            ClearDevice;
```

```
            for i:=start_col to 300 do
            begin
                for j:=0 to 300 do
                begin
                    SetActivePage(back_page);
                    color := getPixel(i+100,j);
                    if color <> 0 then
                    begin
                        iter := 0;
                        while iter < set_end do
                        begin
                            k := Round(a[iter]*i + b[iter]*j +
                                e[iter]);
                            m := Round(c[iter]*i + d[iter]*j +
                                f[iter]);
                            iter := iter + 1;
                            if (k<540) and (m<350) and
                                (k>=-100) and (m>=0) then
                            begin
                                SetActivePage(dis_page);
                                PutPixel(k+100,m,15);
                            end;
                        end;
                    end;
                end;
            end;
            temp := dis_page;
            dis_page := back_page;
            back_page := temp;
        end;
        a[0] := 0;    a[1] := 0.85; a[2] := 0.2;   a[3] := -0.15;
        b[0] := 0;    b[1] := 0.04; b[2] := -0.26; b[3] := 0.28;
        c[0] := 0;    c[1] :=-0.04; c[2] := 0.23;  c[3] := 0.26;
        d[0] := 0.16; d[1] := 0.85; d[2] := 0.22;  d[3] := 0.24;
        e[0] := 0;    e[1] := 0;    e[2] := 0;     e[3] := 0;
        f[0] := 0;    f[1] := 40;   f[2] := 40;    f[3] := 10;
        dis_page := 1; .
        back_page := 0;;
        start_col := -100;
        iterations := 32;
        set_end := 4;
        ch1 := ReadKey;
    end;
end.
```

Figure 20-3. Generation of Sierpinski Triangle by Deterministic Algorithm

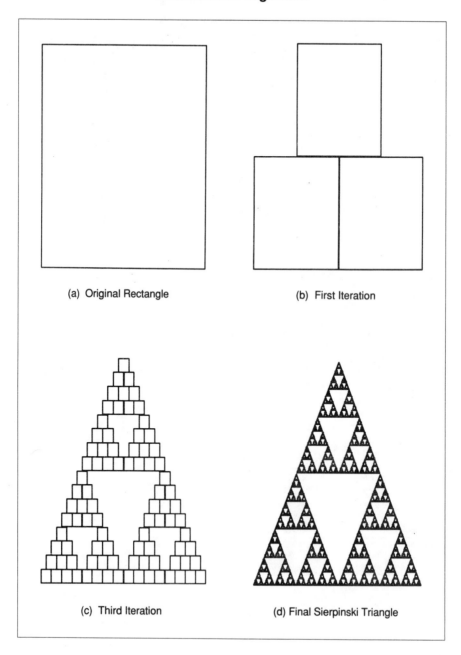

(a) Original Rectangle

(b) First Iteration

(c) Third Iteration

(d) Final Sierpinski Triangle

Generating a Deterministic Fern

The second loop through the deterministic program changes the parameters of the affine transformation so that using the same starting rectangle, we generate a fern leaf.

Generating the fern leaf brings up another problem with the use of the deterministic algorithm. If you aren't careful about the range over which you examine points for transformation, you may get some very weird results. Our Sierpinski triangle worked very well when we transformed over a range of 0 to 200 in both x and y directions. Now look what happens when we use this same range for the fern leaf. The result is shown in Figure 20-4(a).

Somehow, we have generated only one side of each leaflet and have only managed to put one half-leaflet on the left side of the stem. Now change the for loop for the x coordinate to have a range from -100 to +200. You will now get the whole leaf, as shown in Figure 20-4(b). Another drawback of the deterministic algorithm is illustrated when we try to draw the fern. That is the slowness of the process for certain figures. In order to get the pictures in Figure 20-4, we had to go through 32 iterations of the program, and even more wouldn't be out of place. This is quite a lot compared to the few iterations required to produce a good representation of the Sierpinski triangle. There is no simple way to determine in advance how many iterations of the deterministic program are needed to produce a good picture.

Figure 20-4. Generation of Fern Leaf by Deterministic Algorithm

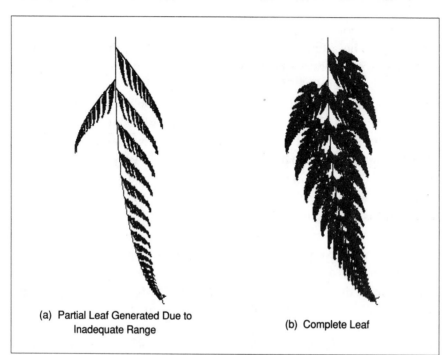

(a) Partial Leaf Generated Due to
Inadequate Range

(b) Complete Leaf

Using the Chaos Algorithm

Figure 20-5 is a table of parameters for several figures that can be generated using the Chaos algorithm. The program for drawing these pictures is listed in Figure 20-6. The program is fairly straightforward. It should be noted that the manner in which one of the affine transformations is randomly selected depends upon comparing a random number between 0 and 1 to the cumulative probabilities of each transformation occurring, so that the last number in each series of comparisons is always one. Note that this program has a background color, a foreground color, x and y scale factors, and x and y offsets included for each picture.

The background color is determined by the number from 0 to 15 that is passed as a parameter in the *SetBkColor* procedure. The foreground color is a number from 0 to 15 passed to the *image_draw* function. The *xscale* and *yscale* numbers determine the size of the picture; you can modify them as you desire to enlarge or reduce the picture. The *xoff-set* and *yoffset* parameters have been selected so that the pictures are nicely centered; you can modify these if you want to reposition the picture on the display. Another thing that you might be interested in doing is to expand the display so that only a portion of the picture appears as an enlarged display. You can do this by changing the scale parameters. The resulting pictures are shown in Figure 20-7.

Figure 20-5. Parameters for Generation of Pictures Using Chaos Algorithm

Figure	a	b	c	d	e	f	p
Sierpinski	0.5	0	0	0.5	0	0	0.33
Triangle	0.5	0	0	0.5	1.0	0	0.66
	0.5	0	0	0.5	0.5	0.5	1.0
Fern Leaf	0	0	0	0.16	0	0	0.01
	0.2	-0.26	0.23	0.22	0	0.2	0.08
	-0.15	0.28	0.26	0.24	0	0.2	0.15
	0.85	0.04	-0.04	0.85	0	0.2	1.0
Tree	0	0	0	0.5	0	0	0.05
	0.1	0	0	0.1	0	0.2	0.20
	0.42	-0.42	0.42	0.2	0	0.2	0.60
	-0.42	0.42	-0.42	0.42	0	0.2	1.0
Cantor	0.333	0	0	0.333	0	0	0.33
Tree	0.333	0	0	0.333	1	0	0.65
	0.667	0	0	0.667	0.5	0.5	1.0

Figure 20-6. Program to Generate Pictures Using Chaos Algorithm

```
program image;

uses CRT,Graph,Fractal;
```

```
var
    GraphDriver,GraphMode,adapt,mode,k,xscale,yscale,xoffset,
        yoffset,pr: integer;
    i: longint;
    a,b,c,d,e,f: array[0..3] of real;
    x,y,newx,j: real;
    p,pk: array[0..3] of real;
    ch1: char;

procedure image_draw(color: integer);

    var
        px,py: integer;

    begin
        x := 0;
        y := 0;
        for i:=1 to 25000 do
        begin
            j := Random;
            if j < p[0] then
                k := 0;
            if (j > p[0]) and (j < p[1]) then
                k := 1;
            if (j > p[1]) and (j < p[2]) then
                k := 2;
            if j > p[2] then
                k := 3;
            newx := (a[k]* x + b[k] * y + e[k]);
            y := (c[k] * x + d[k] * y + f[k]);
            x := newx;
            px := Round(x*xscale + xoffset);
            py := Round(y*yscale + yoffset);
            if (px>=-320) and (px<320) and (py>=-240) and
                (py<240) then
                PutPixel(px+320,175 - ((93*py) div 128),color);
        end;
    end;

begin
    GraphDriver := 4;
    GraphMode := EGAHi;
    InitGraph(graphDriver,GraphMode,'');
    a[0] :=0; a[1] := 0.2; a[2] := -0.15; a[3] := 0.85;
    b[0] := 0; b[1] := -0.26; b[2] := 0.28; b[3] := 0.04;
    c[0] := 0; c[1] := 0.23; c[2] :=0.26; c[3] := -0.04;
    d[0] := 0.16; d[1] := 0.22; d[2] := 0.24; d[3] := 0.85;
    e[0] := 0; e[1] := 0; e[2] := 0; e[3] := 0;
    f[0] := 0; f[1] := 0.2; f[2] := 0.2; f[3] := 0.2;
    p[0] := 0.01; p[1] := 0.08; p[2] := 0.15; p[3] := 1.0;
    xscale := 300;
    yscale := 300;
    xoffset := -50;
    yoffset := -180;
    SetBkColor(1);
    ClearDevice;
```

```
    image_draw(10);
    ch1 := ReadKey;

    a[0] := 0; a[1] := 0.1; a[2] := 0.42; a[3] := 0.42;
    b[0] := 0; b[1] := 0; b[2] := -0.42; b[3] := 0.42;
    c[0] := 0; c[1] := 0; c[2] := 0.42; c[3] := -0.42;
    d[0] := 0.5; d[1] := 0.1; d[2] := 0.42; d[3] := 0.42;
    e[0] := 0; e[1] := 0; e[2] := 0; e[3] := 0;
    f[0] := 0; f[1] := 0.2; f[2] := 0.2; f[3] := 0.2;
    p[0] := 0.05; p[1] := 0.20; p[2] := 0.6; p[3] := 1.0;
    xscale := 750;
    yscale := 750;
    xoffset := 0;
    yoffset := -160;
    SetBkColor(9);
    ClearDevice;
    image_draw(13);
    ch1 := ReadKey;

    a[0] := 0.5; a[1] := 0.5; a[2] := 0.5; a[3] := 0;
    b[0] := 0; b[1] := 0; b[2] := 0; b[3] := 0;
    c[0] := 0; c[1] := 0; c[2] := 0; c[3] := 0;
    d[0] := 0.5; d[1] := 0.5; d[2] := 0.5; d[3] := 0;
    e[0] := 0; e[1] := 1.; e[2] := 0.5; e[3] := 0;
    f[0] := 0; f[1] := 0; f[2] := 0.5; f[3] := 0;
    p[0] := 0.33; p[1] := 0.66; p[2] := 1.0; p[3] := 1.0;
    xscale := 200;
    yscale := 200;
    xoffset := -200;
    yoffset := -160;
    SetBkColor(7);
    ClearDevice;      image_draw(12);
    ch1 := ReadKey;

    a[0] := 0.333; a[1] := 0.333; a[2] := 0.667; a[3] := 0;
    b[0] := 0; b[1] := 0; b[2] := 0; b[3] := 0;
    c[0] := 0; c[1] := 0; c[2] := 0; c[3] := 0;
    d[0] := 0.333; d[1] := 0.333; d[2] := 0.667; d[3] := 0;
    e[0] := 0; e[1] := 1.; e[2] := 0.5; e[3] := 0;
    f[0] := 0; f[1] := 0; f[2] := 0.5; f[3] := 0;
    p[0] := 0.33; p[1] := 0.66; p[2] := 1.0; p[3] := 1.0;
    xscale := 120;
    yscale := 140;
    xoffset := -100;
    yoffset := -160;
    SetBkColor(2);
    ClearDevice;
    image_draw(14);
    ch1:= ReadKey;
end.
```

Figure 20-7. Pictures Generated by Chaos Algorithm

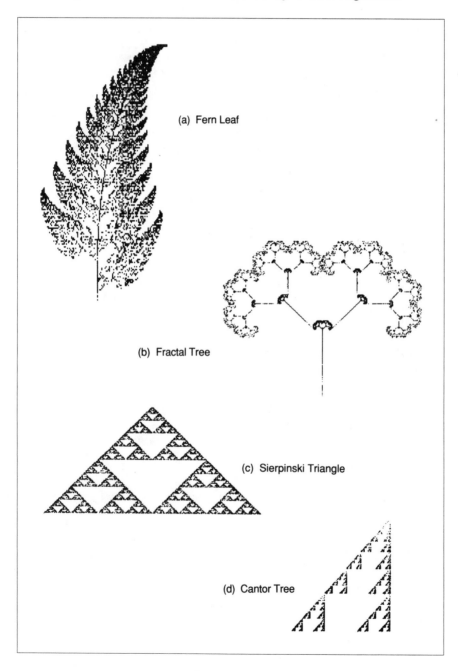

(a) Fern Leaf

(b) Fractal Tree

(c) Sierpinski Triangle

(d) Cantor Tree

The Collage Theorem

Now that you have used the program to generate some pictures with the Chaos algorithm, you may be getting anxious to create some original pictures of your own. I regret to tell you that this is a difficult process which will require some original programming on your part. The theoretical proof that this can be done at all appears in Barnsley's book, where he proves what he calls the *Collage Theorem*. The proof won't help you too much, however, unless you are a mathematician who is expert in set theory.

Basically, what the theorem says is that you can take some picture, or portion of a picture, and by performing affine transformations, end up with smaller (possibly distorted) versions of the original picture which can be placed so that they fill in the original picture without very much overlap, without running very far outside the original picture boundaries, and without leaving much of any blank space. Then by using these same affine transformations in the Chaos algorithm, you can produce a good representation of the original picture. Barnsley has a number of examples.

Now, how do you go about using this theorem? Barnsley doesn't supply any examples that give much clue as to how it is done. (That's not surprising, since he has just formed a company which will do the job for you, if you have a quarter of a million dollars.) You might begin by adding to your computer the capability to display frames captured from a video camera, or you might start with an artist's picture drawn, for example, with PC Paintbrush. Then, you need a program that will perform an affine transformation on the picture or a portion of the picture. It needs to show the result of the transformation and give you the capability to change the transformation parameters until you have the transformed version superimposed just as you want it on part of the original picture. Do this several more times until you have the whole picture covered. Your program then needs to give you the parameters of each of your final affine transformations. You should then be able to insert these in the chaos algorithm pro gram given above, although you may have to modify it to permit it to handle colors in a more sophisticated manner. If this sounds like a very large undertaking, it is. Maybe it will appear in my next book on advanced fractal programming.

The Chaos Algorithm in Three Dimensions

It is possible to make three-dimensional pictures using affine transformations having the form:

$$
\begin{bmatrix} xn \\ yn \\ zn \end{bmatrix} = w \begin{bmatrix} x \\ y \\ z \end{bmatrix} = \begin{bmatrix} a & b & c \\ d & e & f \\ g & h & m \end{bmatrix} \begin{bmatrix} x \\ y \\ y \end{bmatrix} + \begin{bmatrix} n \\ q \\ r \end{bmatrix}
$$

$$
= \begin{bmatrix} ax + by + cz + n \\ dx + ey + fz + q \\ gx + hy + mz + r \end{bmatrix}
$$

Equation 20-2

We don't have the capability to display the resulting picture in three dimensions, so we use the same projection technique that was used for the three-dimensional dragons in Chapter 16.

Figure 20-8 shows the parameters for a three-dimensional fern leaf. The program in Figure 20-9 is a listing of a program to generate the three-dimensional fern leaf four times, using a different shade of green or yellow, and different projection angles for each pass through the loop. The program was set up with the projection angles built in because it takes considerable experimentation to get the right selection of angles to give a pleasing display. The resulting display is shown in Plate 32 (see color section). You can have a lot of fun experimenting with this program by changing offsets and scale factors for each pass through the loop, by modifying the angles, and by increasing the number of passes through the loop to add additional fern leaves.

Figure 20-8. Parameters for Three-Dimensional Fern Leaf

a	b	c	d	e	f	g	h	m	n	q	r	p
0	0	0	0.5	0.18	0	0	0	0	0	0	0	0.01
0.83	0	0	0.5	0.86	0.1	0	-0.12	0.84	0	1.62	0	0.85
0.22	-0.23	0	0.24	0.22	0	0	0	0.32	0	0.82	0	0.92
-0.22	0.23	0	0.24	0.22	0	0	0	0.32	0	0.82	0	1.0

Figure 20-9. Program to Generate Three-Dimensional Pictures Using Chaos Algorithm

```
program image3d;

uses CRT,Graph,Fractal;

const
    alpha: array[0..3] of real = (30,45,15,95);
    beta: array[0..3] of real = (115,105,70,40);
    gamma: array[0..3] of real = (25,70,20,-30);
    rad_per_degree: real = 0.0174533;
    hues: array[0..3] of integer = (2,10,11,14);

var
    GraphDriver,GraphMode,adapt,mode,k,xscale,yscale,xoffset,
        yoffset,pr,index: integer;
    i: longint;
    a,b,c,d,e,f,g,h,m,n,q,r: array[0..3] of real;
    x,y,z,newx,newy,j,ca,cb,cg,sa,sb,sg: real;
    p,pk: array[0..3] of real;
    ch1: char;

procedure image_draw;

    var
        px,py: integer;
        vx,vy: real;

    begin
        x := 0;
        y := 0;
        z := 0;
        for i:=1 to 25000 do
        begin
            j := Random;
            if j < p[0] then
                k := 0;
```

```
            if (j > p[0]) and (j < p[1]) then
                k := 1;
            if (j > p[1]) and (j < p[2]) then
                k := 2;
            if j > p[2] then
                k := 3;
            newx := (a[k]* x + b[k] * y + c[k] * z + n[k]);
            newy := (d[k]* x + e[k] * y + f[k] * z + q[k]);
            z := (g[k] * x + h[k] * y + m[k] * z + r[k]);
            x := newx;
            y := newy;
            vx := x*ca + y*cb + z*cg;
            vy := x*sa + y*sb + z*sg;
            px := Round(vx*xscale + xoffset);
            py := Round(vy*yscale + yoffset);
            if (px>=-320) and (px<320) and (py>=-240) and
                (py<240) then
                PutPixel(px+320,175 - ((93*py) div
                    128),hues[index]);
        end;
    end;

begin
    GraphDriver := 4;
    GraphMode := EGAHi;
    InitGraph(graphDriver,GraphMode,'');
    setEGApalette(0,8);
    setEGApalette(2,2);
    setEGApalette(10,58);
    setEGApalette(11,62);
    setEGApalette(14,26);
    a[0] :=0; a[1] := 0.83; a[2] := 0.22; a[3] := -0.22;
    b[0] := 0; b[1] := 0; b[2] := -0.23; b[3] := 0.23;
    c[0] := 0; c[1] := 0; c[2] :=0; c[3] := 0;
    d[0] := 0; d[1] := 0; d[2] := 0.24; d[3] := 0.24;
    e[0] := 0.18; e[1] := 0.86; e[2] := 0.22; e[3] := 0.22;
    f[0] := 0; f[1] := 0.1; f[2] := 0; f[3] := 0;
    g[0] :=0; g[1] := 0; g[2] := 0; g[3] := 0;
    h[0] := 0; h[1] := -0.12; h[2] := 0; h[3] := 0;
    m[0] := 0; m[1] := 0.84; m[2] :=0.32; m[3] := 0.32;
    n[0] := 0; n[1] := 0; n[2] := 0; n[3] := 0;
    q[0] := 0; q[1] := 1.62; q[2] := 0.82; q[3] := 0.82;
    r[0] := 0; r[1] := 0; r[2] := 0; r[3] := 0;
    p[0] := 0.01; p[1] := 0.85; p[2] := 0.92; p[3] := 1.0;
    xscale := 40;
    yscale := 50;
    xoffset := 60;
    yoffset := -180;
    SetBkColor(0);
    ClearDevice;
    for index:=0 to 3 do
```

```
begin
    ca := cos(alpha[index]*0.0174533);
    cb := cos(beta[index]*0.0174533);
    cg := cos(gamma[index]*0.0174533);
    sa := sin(alpha[index]*0.0174533);
    sb := sin(beta[index]*0.0174533);
    sg := sin(gamma[index]*0.0174533);
    image_draw;
end;
ch1 := ReadKey;
end.
```

Appendix A

Hardware Requirements

With the wide variety of PC clones that are now available, including some that use 80286 and 80386 microprocessors at speeds far higher than anything envisioned by IBM, there is no telling when some strange glitch is going to wreak havoc with your program. To attempt to minimize that kind of problem, I have attempted to run most of the programs described in this book on two different systems. One consists of the following:

Motherboard: Bullet-286E from Wave Mate, Inc., Torrance, CA. This board is a drop-in replacement for the PC XT motherboard. It uses an 80286 and has 1 megabyte of on-board 0 wait state memory. The memory from 640K to 1 MByte is used as a hard disk cache. An 80287-6 math coprocessor was used.

Floppy Disk Controller: MCT-FDC-1.2 from JDR Microdevices, Los Gatos, CA. This board supports 360K or 1.2M disk drives.

Floppy Disk Drives: 1 Fujitsu 1.2 Megabyte disk drive. 1 Mitsubishi 360Kbyte floppy disk drive.

I/O Board: MCT-IO from JDR Microdevices.

Hard Disk Controller: MCT-RLL from JDR Microdevices.

Hard Disk Drive: LaPine LT300 30 Megabyte from Advanced Computer Products, Santa Ana, CA.

Keyboard: Surplus Honeywell keyboard from B. G. Micro, Dallas, TX.

EGA Card: MCT-EGA from JDR Microdevices.

EGA Monitor: Casper EGA Monitor from JDR Microdevices.

The second system consists of:

Motherboard: DTK Baby AT with 80286. Speeds 8 and 12 MHz. 1 Mbyte memory on-board, running with 1 wait state, from U. S. Turbo Systems, South El Monte, CA.

Floppy Disk Controller: Techway Multi Driver II Adapter from Gems Computers, Inc., Santa Clara, CA.

Floppy Disk Drives: 1 Fujitsu 1.2 M from Gems Computer; 1 360K Qume from Jade Computer, Hawthorne, CA.; 1 Sony 1.44 M 3 1/2 inch Drive

VGA Card: Vega VGA from Video Seven, Fremont, CA.

VGA Monitor: NEC Multisync Plus.

Keyboard: KeyCat keyboard (a 101 key keyboard with a built-in trackball) from Dexxa International, Burlingame, CA.

Some cards were occasionally swapped between the two systems described above. In addition, text was printed out on a Hewlett-Packard LaserJet IIP Printer, and color pictures were printed out on a Hewlett-Packard PaintJet Printer.

Programs that work satisfactorily on systems as different as these two should have a good chance of working on your system too. If you encounter problems, try to identify how your system differs from the two described above.

Display Considerations

I highly recommend that if you are going to get into fractals seriously, you get an EGA or VGA display. Much of the beauty and power of fractal displays comes from manipulating the colors to match the conceptions of the programmer/artist. You will find a number of black-and-white illustrations of fractal curve results throughout this book. If you must work with a Hercules Graphics Card or compatible monochrome display, you can reproduce these illustrations. You will find full details on how to use the Hercules Graphics mode in my book *Graphics Programming in C*.

All of the color illustrations in the book were done using color mode 16 (which is high-resolution, 16 colors, and is common to the EGA and VGA). You can duplicate all of the color programs using this mode, and then if you have a VGA, you can go to mode 18 and with a few minor changes obtain greater vertical resolution. The VGA also permits you to select the 16 color palettes used in modes 16 or 18 from 256K shades of color instead of the 64 available with the EGA. In most cases, however, you are going to find that the EGA gives you plenty of color capability.

I don't recommend using the CGA color monitor for fractal displays. The resolution is just not adequate for the detail needed to produce interesting fractals. If you have a VGA, mode 19 has the same resolution as the CGA, but permits you to display 256 different colors simultaneously. You're welcome to experiment with that, but I think you will find that the additional picture detail is much to be preferred over the additional color shades.

One word of caution. EGA and VGA adapters that are not made by IBM are not always completely compatible with IBM standards. Thus, some of the procedures and functions in this book that directly access the EGA or VGA board may not work properly with off-beat graphics adapter cards. If you are having trouble, the best thing to do is select one of the programs that uses only the Turbo Pascal graphics modes as a test case. If you can't run this, try a very simple program that simply draws a line using Turbo Pascal graphics. Unfortunately, I can't help you if this doesn't work.

Processor Speed

As you get into the more complicated fractal displays, such as the Mandelbrot and Julia sets, you are going to find that processor speed becomes more and more of a problem. Even on a fairly fast machine, many of the more interesting displays take days to generate. How can this be resolved? First, code needs to be optimized so that the loops that are iterated the most times have as simple procedures as possible. Next, if you have one of the fast 286 clones, makes sure you have fast enough memory so that you can run the programs at the highest available clock speed and preferably with no wait states. Finally, get a math coprocessor chip. Steve Ciarcia says in the December 1988 issue of *Byte* that tests he ran show that the coprocessor runs the Mandelbrot set program eight times faster than with the math routines emulated in the C language. Similar speed increases should also apply to Turbo Pascal. Turbo Pascal is set up to automatically make use of a math coprocessor if one is available in your system.

If you do not have a coprocessor, an attractive alternative might appear to be generating your own assembly language routines to perform the special math processing that uses most of the time in generating the Mandelbrot sets and similar programs. This, however, turns out not only to be difficult, but not to be as timesaving as one might think. The programmers who wrote the math routines for Turbo

Pascal spent a lot of time making these routines as fast and efficient as possible. You are not likely to come up with anything faster for a particular operation, and if you try to save by reducing the precision of the routines, you will probably find that your display looks inferior compared to ones generated by more conventional means. The remaining time saving that you might achieve is by holding results in registers and doing all of the iterations without transferring to memory, but this will require a lot of register manipulation for 8086 or 80286 machines, which probably isn't worth it.

There are a couple of exceptions to the above. If you have an 80386-based machine, the internal 32 bit registers provide sufficient precision to perform good fractal computations. H. W. Stockman, in the Sept.–Oct. 1988 issue of *Micro Cornucopia* describes how to write 386 assembly language code to compute the Mandelbrot set directly on the microprocessor's internal registers. He claims a speed-up of 100 using this technique. Once you understand his assembly language code, you should be able to adapt it for generating fast Julia sets or dragons or phoenixes. I have recently come across an interesting program called FRACTINT which contains programs for generating lots of different fractals on any PC that uses an 80286 or 80386. This program uses integer arithmetic for many types of fractals and is very fast. It is available free on many bulletin boards. You can get further information from:

Timothy Wegner
1955 Portsmouth
Houston, Texas 77098

The only trouble with FRACTINT is that it does so much with so many different kinds of fractals that it becomes hard to understand all of the techniques and algorithms that are used.

Another approach was used by Steve Ciarcia in the October, November, and December 1988 issues of *Byte*. Steve built a parallel processor which uses 64 Intel 8051 microprocessors in parallel in a special Mandelbrot set generating computer. The cost of this monster, in addition to its PC driver, runs around $6000. Steve found that it about matched the AT with coprocessor when he used 15 of the 8051s and was many times faster with 64 parallel processors. And, of course, there is nothing to prevent you from stopping there. In the extreme, you could have one 8051 for each pixel on the display, all working simultaneously and limited only by the time required to dump the results to the display. If you're going to take either of the above routes, you need some real dedication; in most cases, an AT type machine with an EGA display and a math coprocessor will generate all of the fractal displays you will ever need or want without delays that are objectionably long.

Appendix B

Format for PCX Files

The ZSoft .PCX file format begins with a 128 byte header, the contents of which are shown below:

Header Data for .PCX Screen File

Byte	Size (Bytes)	Name	Description
0	1	Password	0AH designates ZSoft.PCX files.
1	1	Version	versions of PC Paintbrush 0 = vers 2.5 2 = vers 2.8 w/palette info 3 = vers 2.8 w/o palette information 5 = vers 3.0
2	1	Encoding	Encoding scheme used 1 = .PCX run length encoding.
3	1	Bits per pixel	No. of bits required to store data for 1 pixel from 1 plane. 1 = EGA, VGA or Hercules 2 = CGA
4	8	Window dimensions	4 integers (2 bytes each) giving top left and bottom right corners of display in order x1, y1, x2, y2.
12	2	Horizontal Resolution	Horizontal resolution of display device

Byte	Size	Name	Description
			320 = CGA
			720 = Hercules
14	2	Vertical Resolution	Vertical resolution of display device (lines)
			480 = VGA
			350 = EGA
			200 = CGA
			348 = Hercules
16	48	Color Map	Information on color palette settings. See following figures for details.
64	1	Reserved	
65	1	Number of planes	Number of color planes in the original image
			1 = CGA, Hercules
			4 = EGA, VGA
66	2	Bytes per line	Number of bytes per scan line in the image
68	2	Palette Description Information	How to interpret palette.
			1 = color/monochrome
			2 = grayscale
70	16	Picture limits	Four floating point numbers giving the bounds for the set computation. The order is XMax, XMin, YMax, YMin. This section is not used in original ZSoft format.
86	8	Iteration	Values for Pval and Qval used as parameters for fractal computation. This section is not used in original ZSoft format.
94	32	Not used	Fill to end of header block.

Except for the color map, most of the header contents are self-evident. The floating point numbers for *XMax*, *XMin*, *YMax*, *YMin*, *Pval*, and

Qval are not part of the original ZSoft format. They are needed when we save files for Mandelbrot sets and similar displays to preserve data required to define the figure. They are defined as *single* numbers so as to match the space requirements that are used for the same file in *Fractal Programming in C*. Thus, files can be used with programs from that book or this one interchangeably. Since that part of the ZSoft header which contains *XMax, XMin, YMax, YMin, Pval,* and *Qval* is normally empty, using any other ZSoft-compatible program with these files should not present a problem. If the file is read by one of the programs described in the later chapters of this book, the values will be extracted and used as needed. If the file is read by another program, the values will be ignored.

The contents of a palette register for the EGA color system is shown below. Six bits are used, with two each for the primary colors red, green, and blue. The capital letters represent colors of 75% amplitude; the small letters colors of 25% amplitude. Thus, for each of the primary colors, four levels are available: 0 (none of that color), 25% amplitude, 75% amplitude, and 100% amplitude (both capital and small letter bits are one). The color map in the file header contains 16 sets of triples, one for each EGA palette. For the first byte of a triple, the values of the capital and small letter position for red are extracted and combined to produce a number from one to three. This number is multiplied by 85 and stored in the header.

The same procedure takes place for the second byte of the triple for green, and the third byte for blue. The process is repeated sixteen times, once for each palette. Note that when we set the palette registers on the EGA we are setting a write-only register, so that we can never recover the contents if we want to know later what the setting was. Consequently, our *setEGApalette* function saves the palette register information in a global array *PALETTE* array [0..15] of integer. It is this data that we use to write the color map in the header when we are saving a screen.

The VGA is quite different in the way that it handles colors. With the VGA, each palette register contains the number of a color register. The color register contains six bits, permitting 64 shades of each color. With the VGA we can read the six-bit value for each of red, green, and blue, information in the palette registers and in the color registers.

To create the color map, we read each palette register and then go and read the color register pointed to by that palette register. We then multiply the red, green, and blue values by four and store the results in the triple associated with that palette.

Note that when we restore a screen, we may not assign a color value to the same color register that it was obtained from originally, and that the palette registers may not select the same color registers. However, the net result is the same, because each palette register points to a color register that contains the same color information that was contained in the original screen.

The VGA also has a color mode in which 256 different colors may be displayed simultaneously. Since this is at a lower resolution, (the same resolution as the CGA display) we won't be using it in our screen saving and restoring functions. For reference, the format is the same as used to display the 16 color palette, but due to the 256 colors, the palette information is much longer. It is appended at the end of the .PCX file. To access this information, you must first ascertain that the version number data in the header (byte 1) is 5 (version 3.0). Then read to the end of the file and count back 769 bytes. If the value in this byte position is 0CH (12 decimal), the succeeding information is 256 color palette data.

Data is read from the screen, horizontally from left to right, starting at the pixel position for the upper left corner. For EGA and VGA, which have multiple memory planes, a line is read of the color red (to the end of the window boundary), then the green information for the same line is read, and finally the blue.

The functions that are developed in Chapter 3 only work if the horizontal pixel boundaries are at a byte interface (the column number must be divisible by eight.) Data is run length encoded in the following manner. If the byte is unlike the ones on either side of it, and if its two most significant bits are not 11, it is written to the file. Otherwise a count is made of the number of like bytes (up to 63) and this count is ANDed with C0H. The result is written to the file, followed by the value of the byte. If there are more that 63 successive, like bytes, the count for 63 and the byte are written, and then the count begins all over again. (Note that the case for a singular byte having the two most significant bits 1 is handled by writing a count of one followed by the byte value.)

Contents of EGA Palette Register

Byte 7	Byte 6	Byte 5	Byte 4	Byte 3	Byte 2	Byte 1	Byte 0
		r 25%	g 25%	b 25%	R 75%	G 75%	B 75%

Contents of .PCX File Color Map

Byte	Palette	Color	Description
16	0	Red	For the EGA, the values of
17	0	Green	color of each byte of each
18	0	Blue	triple are:
19	1	Red	00H to 54H = 0%
20	1	Green	55H to A9H = 25%
21	1	Blue	AAH to FEH = 75%
22	2	Red	For the VGA, the value of each
23	2	Green	byte is the value of the six-
24	2	Blue	bit color value from the color
25	3	Red	register pointed to by the
26	3	Green	appropriate palette register
27	3	Blue	multiplied by four.

Byte	Palette	Color	Description
28	4	Red	
29	4	Green	
30	4	Blue	
31	5	Red	
32	5	Green	
33	5	Blue	
34	6	Red	
35	6	Green	
36	6	Blue	
37	7	Red	
38	7	Green	
39	7	Blue	
40	8	Red	
41	8	Green	
42	8	Blue	
43	9	Red	
44	9	Green	
45	9	Blue	
46	10	Red	
47	10	Green	
48	10	Blue	
49	11	Red	
50	11	Green	
51	11	Blue	
52	12	Red	
53	12	Green	
54	12	Blue	
55	13	Red	
56	13	Green	
57	13	Blue	
58	14	Red	
59	14	Green	
60	14	Blue	
61	15	Red	
62	15	Green	
63	15	Blue	

About the Author

Dr. Roger T. Stevens graduated from California Western University, in Santa Ana, California, with a PhD in Electrical Engineering, but not before attaining degrees from three other universities, including a Masters in Engineering in Systems Engineering at Virginia Polytechnic Institute in Blacksburg, Virginia; a Master of Arts in Mathematics at Boston University in Boston; and a Bachelor of Arts in English at Union College in Schenectady, New York.

Dr. Stevens's engineering career spans more than 30 years. He is currently a member of the technical staff of the MITRE Corporation, Bedford, Massachusetts, in charge of system engineering support and development for the Defense Test and Evaluation Suport Agency.

He has written numerous articles for technical magazines and journals, and is the author of *Operational Test and Evaluation: A Systems Engineering Process* (John Wiley, 1978), *Graphics Programming in C* (M&T Books, 1988), and *Fractal Programming in C* (M&T Books, 1989). Dr. Stevens resides with his wife, Barbara, in Albuquerque, New Mexico.

Bibliography

Ackerman, M., "Hilbert Curves Made Simple," *Byte*, June 1986, pp. 137–138.

Barnsley, M. F., and Sloan, A. D., "A Better Way to Compress Images," *Byte*, January 1988, pp. 215–223.

Barnsley, Michael, *Fractals Everywhere*, Boston, 1988, Academic Press, Inc.

Barnsley, M. and Demko, S., "Iterated Function Systems and the Global Construction of Fractals," *Proceedings R. Soc. Lond*, A 399, 1985, pp. 243–275.

Batty, M., and Longley, P. A., "Fractal–Based Description of Urban Form," *Environment and Planning B: Planning and Design*, Vol. 14, 1987, pp. 123–134.

Batty, M., "Fractals—Geometry Between Dimensions," *New Scientist*, April 1985, pp. 31–35.

Batty, Michael, *Microcomputer Graphics*, London, 1987, Chapman and Hall.

Batty, M. and Longley, P. A., "Urban Shapes as Fractals," *Area*, 19.3, 1987, pp. 215–221.

Bouville, C., "Bounding Ellipsoids for Ray-Fractal Intersection," *SIGGRAPH '85*, Vol. 19, No. 3, pp. 45–52.

Butz, A. R., "Alternative Algorithm for Hilbert's Space-Filling Curve," *IEEE Transactions on Computers*, Vol. C-20, No. 4, April 1971, pp. 424–426.

Collet, P., Eckmann, J.-P., and Lanford, O. E., "Universal Properties of Maps on an Interval," *Communications in Mathematical Physics*, 76, 1980, pp. 211–254.

Demko, S., Hodges, L., and Naylor, B., "Construction of Fractal Objects with Iterated Function Systems," *SIGGRAPH '85*, Vol. 19, No. 3, pp.271–278.

Dewdney, A. K., "Computer Recreations," *Scientific American*, September 1986, pp. 140–145.

Feigenbaum, M. J., "Quantitative Universality for a Class of Nonlinear Transformations," *Journal of Statistical Physics*, Vol. 19, No. 1, January 1978, pp. 25–52.

Fogg, L., "Drawing the Mandelbrot and Julia Sets," *Micro Cornucopia*, #39, January–February 1988, pp. 6–9.

Fogg, L., "Fractal Miscellany," *Micro Cornucopia*, #43, September–October 1988, pp. 48–51.

Fogg, L., "Introduction to Fractals," *Micro Cornucopia*, #33, December–January 1987, pp. 36–40.

Fournier, A., Fussell, D., and Carpenter, L., "Computer Rendering of Stochastic Models," *Communications of the ACM*, Vol. 25, No. 6, June 1982, pp. 371–384.

Gleick, James, *Chaos: Making a New Science*, New York, 1987, Viking.

Hirst, K. E., "The Apollonian Packing of Circles," *Journal London Math Society*, 42, 1967, pp. 281–291.

Land, B. R., "Dragon," *Byte*, April 1986, pp. 137–138.

Lanford, O. E., "A Computer-Assisted Proof of the Feigenbaum Conjectures," *Bulletin of the American Mathematical Society*, Vol. 6, No. 3, May 1982, pp. 427–434.

Li, T. Y., and Yorke, J. A., "Period Three Implies Chaos," *American Mathematical Monthly*, 82, December 1975., pp. 985–992.

Lorenz, E. N., "Deterministic Nonperiodic Flow," *Journal of the Atmospheric Sciences*, Vol. 20, March 1963, pp. 130–141.

Mandelbrot, B., "Comment on Computer Rendering of Fractal Stochastic Models," *Communications of the ACM*, Vol. 25, No. 8, pp. 581–584.

Mandelbrot, B., *The Fractal Geometry of Nature*, New York, 1983, W. H. Freeman and Company.

Mastin, G. A., Watterberg, P. A., and Mareda, J. F., "Fourier Synthesis of Ocean Scenes," *IEEE Computer Graphics and Applications*, March 1987, pp. 16–24.

Mastin, G. A., "From Research into Art," *IEEE Computer Graphics and Applications*, March 1987, pp. 5–8.

Norton, A., "Generation and Display of Geometric Fractals in 3-D," *Computer Graphics*, Vol. 16, No. 3, July 1982, pp. 61–67.

Patrick, E. A., Anderson, D. R., and Bechtel, F. K., "Mapping Multidimensional Space to One Dimension for Computer Output Display," *IEEE Transactions on Computers*, Vol. C-17, No. 10, October 1968, pp. 949–953.

Peitgen, H. O., and Richter, P. H., *The Beauty of Fractals*, Berlin, 1986, Springer-Verlag.

Peitgen, H. O., Saupe, D., *The Science of Fractal Images*, Berlin, 1988, Springer-Verlag.

Pickover, C. A., "A Note on Rendering 3-D Strange Attractors," *Computers and Graphics*, Vol. 12, No. 2, 1988, pp. 263–267.

Preston, F. H., Lehar, A. F., and Stevens, R. J., Compressing, ordering and displaying image data," *SPIE Vol 359 Applications of Image Processing IV*, 1982, pp. 302–308.

Robinson, F., "Plotting the Mandelbrot Set with the BGI," *Turbo Techix*, May–June 1988, pp. 28–35.

Sorensen, P., "Fractals," *Byte*, September 1984, pp. 157–171.

Stockman, H. W., "Fast Fractals," *Micro Cornucopia*, #43, September–October 1988, pp. 22–29.

Ushiki, S., "Phoenix," *IEEE Transactions on Circuits and Systems*, Vol. 35, No. 7, July 1988, pp. 788–789.

Zorpette, G., "Fractals: Not Just Another Pretty Picture," *IEEE Spectrum*, October 1988, pp. 29–31.

More Turbo Pascal Tools & Resources from M&T Books

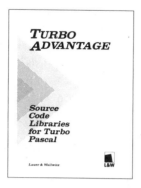

Turbo Advantage
Source Code Libraries for Turbo Pascal
by Lauer & Wallwitz

The *Turbo Advantage* library contains more than 220 routines complete with source code, sample programs, and documentation. Routines are organized and documented in categories such as file management, MS-DOS support, string operations, data compression, and more! The detailed manual includes a desciption of each routine, an explanation of the methods used, and the calling sequence.

Manual and Disk (MS-DOS)   *Item #26-7*   *$29.95*

Turbo Advantage Display
Form Generator for Turbo Pascal
by Lauer & Wallwitz

Turbo Advantage Display form generator makes it easy to design and process forms. This package includes an easy-to-use form processor, 30 Turbo Pascal procedures and functions to facilitate linking created forms to your program, and full source code and documentation. Some of the *Turbo Advantage* routines are necessary to compile *Turbo Advantage Display*.

Manual and Disk (MS-DOS)   *Item #28-3*   *$39.95*

More Turbo Pascal Tools and Resources ...

Turbo Advantage Complex
Complex Number Routines for Turbo Pascal

by Lauer & Wallwitz

Turbo Advantage Complex provides procedures for performing all the arithmetic operations and necessary real functions with comlex numbers. Each procedure is based on predefined constants and types. *Turbo Advantage Complex* demonstrates the use of these procedures in routines for vector and matrix calculation with complex numbers and variables, and calculations of convolution and correlation functions. Source code and documentation are included. Some of the routines are most effectively used with routines contained in *Turbo Advantage*.

Manual and Disk (MS-DOS)   *Item #27-5*   *$39.95*

- -

To Order: Return this form with your payment to **M&T Books**, 501 Galveston Drive, Redwood City, CA 94063 or **CALL TOLL-FREE 1-800-533-4372** (in California, call 1-800-356-2002).

☐ **YES!** Please send me the following: ☐ Check enclosed, payable to **M&T Books**.

Item#	Description	Qty	Price

Charge my ☐ Visa ☐ MC ☐ AmEx

Card No. _____ Exp. Date_____

Signature_____

Name_____

Address _____

Subtotal _____

CA residents add sales tax —%_____

Add $2.99 per item for shipping

and handling_____

TOTAL _____

City _____

State _____ Zip_____

Note: Prices subject to change without notice. Disks may be returned for exchange only if damaged in transit.